Bachelor
Colorado

*History of a San Juan Mining
Ghost Town*

By

Charles A. Harbert

WESTERN REFLECTIONS PUBLISHING COMPANY®

ISBN : 978-1-937851-26-2

Library of Congress Control Number: 2017933037

First Edition
Printed in the United States of America

Western Reflection Publishing Company
Lake City, Colorado
www.westernreflectionspublishing.com

Dedication

It is a pleasure to dedicate this book to Richard (Dick) Huston (1931–2016), a good friend and fellow lover of Creede history. Dick wrote *A Silver Camp Called Creede: A Century of Mining* in 2005, followed by *A Gold Camp Called Summitville* in 2012. His book on Creede is the definitive reference book on mining in Creede Camp. It is a must read for anyone interested in the history of Creede.

I met Dick in 2004 in Creede, Colorado, when he was doing research on his first book. We continued our friendship in Tucson, Arizona, where my wife, Kay, and I had lunch with Dick and his lovely wife, Kathy, every Sunday.

Dick grew up in Monte Vista, Colorado, and graduated from the Colorado School of Mines in 1953. He worked in the Commodore Mine as a mucker and miner during summer vacations and gained an appreciation for what it meant to be an underground miner. He had an illustrious career working on engineering and construction projects in the mining industry worldwide.

Dick had a broad and deep knowledge of Creede and all aspects of mining in the area. He had a great sense of humor and was quick to tell stories about his experiences in Creede, the colorful people he met, and life in general. He was a frequent visitor to Creede before and after he wrote his book. As with many miners, breathing problems kept him from coming to Creede in his final years. Regardless, his heart is still in Creede! He will be missed.

Contents

Preface   i
Acknowledgments   v
Chapter 1   Creede Caldera and *Ancient Lake Creede*   1
Chapter 2   Beginnings   8
Chapter 3   Birth of a New Mining Town   18
Chapter 4   Incorporation   29
Chapter 5   Boomtown   42
Chapter 6   Life in Bachelor   74
Chapter 7   A Town in Decline   102
Chapter 8   Memories of Bachelor   123
Chapter 9   People of Bachelor   134
Epilogue   142
Appendix 1   Ordinances of the Town of Bachelor   146
Appendix 2   *Colorado Business Directory* Entries
                    for Bachelor   165
Notes   178
Bibliography   189
Index   193

PREFACE

The phrase "Capture the Shadow Ere the Substance Fades" was a term that nineteenth century photographers used to encourage people to take postmortem photographs of their loved ones. The hope was that a photograph would preserve their memory. This principle of historical preservation is part of my motivation to write books about and collect old photographs of Creede and Bachelor.

In writing this book, I am capturing the shadow of Bachelor in words and photographs before it fades completely. I was also motivated by the fact that a book on Bachelor had not been written and by an innate curiosity. What better way to learn about Bachelor than to write a book about this mining ghost town?

Admittedly, it would have been better if someone had written a book before 1930 while parts of the town were still standing, you could interview people who lived or had lived there, and newspapers from the town were still accessible. But that didn't happen. So, I decided to do it now. I hope you, the reader, believe it was worth the effort.

I uncovered several surprises in researching and writing the book. The first is that many contemporaries considered Bachelor to be a wilder town than Creede. Men from Creede frequently traveled to Bachelor to have a good time. Poker Alice, of Creede fame, spoke highly of the gambling opportunities in Bachelor. On a per capita basis, there were probably more violent episodes in Bachelor than there were in Creede. Second, I learned that Ed O'Kelley (the murderer of Bob Ford) was one of Bachelor's founders and played an important role in the town's incorporation. I was able to identify for the first time the name of O'Kelley's accomplice, who was also one of Bachelor's founders. I was pleased to find several first-hand stories of Bachelor that are incorporated in the book. Finally, I was surprised to learn of the close relationship between the town of Bachelor and the Last Chance Mine. Their fates were inter-twined from the beginning and they frequently shared important social events.

Each time I drive by the open meadow that was once the town of Bachelor, I marvel at how little is left. Comparison of the photographs in figures 1 and 2 dramatically illustrates this point. At its peak in 1893, the population of the booming town was between 1,200 and 1,500. *Teller Topics*, the Bachelor newspaper, boasted in 1892 that the town would become the business center of Creede Camp and perhaps all of southwestern Colorado. For various reasons, explained later, that never happened.

In preparation for the book, I walked around the Bachelor town site several times and found it to have a mystical quality. While it is difficult to find anything that reminds us that this was once a booming town, it is easy to be transported back into the 1890s when you walk through the town site. I could easily imagine walking on Main Street and talking with Gustav Hoffman (first mayor), Will Wood (postmaster), Edith Oberg Lundy (postmistress), and Ed O'Kelley (Bob Ford's killer). It must have been an exciting time.

It is easy to see why John MacKenzie and Sam Coffin founded the town here. The location was close to the large silver mines on the Amethyst Vein, had natural springs, and commanded a great view of the Upper Rio Grande Valley. With its southern exposure, it was bathed in sunlight in summer and winter. Ute Indians almost certainly took advantage of the meadow as a summer camp—long before MacKenzie, Coffin, and Nicholas Creede walked on Bachelor Mountain.

The incline from the bottom to the top of the town is a bit of a challenge to negotiate, so it is easy to understand why Horace Wheeler, as a boy of approximately age ten, was jealous of the Cummings boys because he and his sister had to carry water up the hill each day while the neighbor boys had a burro to carry their water up the hill.

When I wrote *Creede, Colorado History: Insights and views through postcards and photographs*, I was surprised at how little information had been published about Bachelor. Both Nolie Mumey's *Creede: The History of a Silver Mining Town* and Dick Huston's *A Silver Camp Called Creede* have a few pages on the town but nothing substantial.

In my position as a volunteer at the Creede Historical Society Museum, I have met several people with ancestors who lived in Bachelor. The discussions were delightful, but we had little published information to give them. I wrote this book because I hope it fills an information void for people who want to learn more about the town.

There are many stories buried under the sod at the placid meadow. I have uncovered a few of them here. Despite years of searching, I know there are many more stories I did not find.

I wrestled with whether to include the geologic origins of the Creede area in an Appendix or Chapter 1. To me, this information is critical to a full understanding of how silver was deposited in the mountains and the Upper Rio Grande Valley was created. As a result, I included it as Chapter 1. Parts of this chapter and Chapter 2 were extracted from my Creede book published in 2010.

In Chapter 2, I include a discussion of the Stony Pass Route and the discovery of the Amethyst Vein because they are integral to Bachelor's founding. This chapter introduces John MacKenzie, one of the earliest prospectors and the "Father of Bachelor," as well as the discovery of the mines on the Amethyst Vein. Indeed, the rationale for the founding of Bachelor was its proximity to these mines.

I included the town ordinances as Appendix 1 because I thought it was useful for readers to understand the legal principles by which the town operated. They also demonstrate the widespread optimism among Bachelor's leaders that the new town would grow into a major city that would need them. Appendix 2 provides Business Directory listings. This gives a flavor of the mix of businesses in the town and also illustrates how the number of businesses decreased along with the population, beginning in the late 1890s. I considered including U. S. Census data for 1900 and 1910 but decided against it for the sake of brevity. I typed up the census information and have made it available for reference at the Creede Historical Society Library in Creede, Colorado. The census data are organized, first, as collected by the census taker to allow the reader to identify family or living units and, second, in alphabetical order to aid in locating people with a specific last name.

Finally, the story of the birth and demise of Bachelor is beautifully captured in a poem by Jane Morton that is included in the Epilogue.

Fig. 1. Photograph of the town of Bachelor on July 24, 1895. Unlike other photographs of the town, it shows the entire expanse from south to north. The hills surrounding the town have been denuded of trees. Made from a glass negative in the Harbert Archives.

Fig. 2. Photograph by Bob Seago, taken in summer 2015 from the same position as the 1895 photograph in figure 1. It shows how little is left of the town and that aspen trees have grown back in around the periphery. The structure to the right is a modern home built on private property.

ACKNOWLEDGMENTS

Many people and institutions helped me assemble the reference material and photographs necessary to write this book. I owe them a great debt of gratitude.

Jan Jacobs and Bob Seago supplied a large number of photographs from the Creede Historical Society Archives that were used in the book. They also provided helpful comments on the final draft of the book. Jan, Bob, and Carol Pierce have labored steadily over the past several years to organize and catalog the Historical Society's extensive photo library. During preparation of the manuscript, they frequently sent me new photographs they uncovered in the library files. Thanks also to Jan and Bob for helpful comments on the last draft of the book.

Johanna Gray was extremely helpful in getting me started on the project. She searched the library files and uncovered a wealth of unpublished information on Bachelor and families who lived there. She continued to be responsive to requests while I was writing the manuscript.

In 2015 I acquired a glass negative of a rare 1895 photograph of Bachelor (fig. 1) showing the entire expanse of the town. I thank Bob Seago for climbing Bulldog Mountain with me to identify the exact location from which the original photograph was taken. It was exhilarating to stand in the same spot the original photographer had stood 120 years earlier. Bob took the current-day photograph (fig. 2) matching the one taken from the same location in 1895. The 1895 and 2015 photographs illustrate how quickly a booming town can return to its natural state.

Grant Houston, editor of Lake City's *Silver World* (newspaper), arranged for me to view and copy the original version of the request for incorporation by Bachelor residents and the official vote for incorporation that were stored at the Hinsdale County Museum in Lake City. He also allowed me to photograph an original copy of the town plat. I was delighted to find that the museum has one of the manual water pumps used in Bachelor. More recently, Houston located the official Bachelor seal stamp tool in the museum. Photographs of the

Fig. 3. Bob Seago on Bulldog Mountain the day he and I climbed the mountain to photograph the Bachelor town site from the exact spot another photographer took a photograph in July 1895 (see figs. 1 and 2). The outcropping also has a spectacular view of Creede.

incorporation documents, town plat, water pump, and seal stamp tool are depicted in this book.

The staffs at History Colorado and the Denver Public Library were extremely helpful in guiding me to key reference material. Particular thanks to Sara Gilmor, reference librarian at the Stephen H. Hart Library and Research Center, who guided me to 1892 copies of *Teller Topics* that provided valuable resource material for the book. It was a pleasure to work with her and other librarians, face-to-face and through COsearch, the library's online help service.

Colorado Historic Newspaper Collection, an online resource to access images of old newspapers suggested to me by Johanna Gray was extremely helpful. I used it constantly to search *Creede Candle* newspaper articles and verify references. It would have been extremely difficult to conduct research without this valuable resource.

Thanks to Larry and Irene Gardanier, descendants of Edith Eleanor Oberg Lundy Gardanier, who grew up in Bachelor in the 1890s and later became postmistress of the town. She married Sutter A. Gardanier in

1909 and moved to Creede. I had the pleasure of meeting Larry and Irene when I was volunteering at the Creede Historical Society Museum. They have been generous in sharing information and photographs of Bachelor and Edith, some of which are included in the book.

In an interview in the fall of 2015, Delma and Bill Dooley provided interesting stories about Bachelor that I have used in the book. Delma let me photograph the hinged key found near the Bachelor Jail (see fig. 59). I followed up with a telephone conversation with Norah Dooley Korn, who described how she and her mother, Alma Lucille Wintz Dooley, found the key in front of the decaying jail while on a picnic at the Bachelor town site. The key fit the lock but would not turn because the lock was rusted.

Special thanks to the people who recorded their recollections of Bachelor in writing and made them available at the Creede Historical Society Library. By doing this, they left a lasting legacy for their families and the people of Mineral County. In particular, Harold French Wheeler, Edwin Lewis Bennett, Don LaFont, and others took the time to put their memories into words. Excerpts from these references form an important part of the book. Special thanks also to Rena Bresser, who sent me an updated copy of the *Family History of Harold Wheeler and Muriel LeZotte*, and to Susan Weston, Mr. French's granddaughter, who updated and edited his memoirs shortly before he died.

Clyde Dooley graciously provided information on Andy Dooley, who lived in Bachelor briefly in the early 1900s before he moved to Creede. Lewis J. (Bud) Wood provided information about and photographs of Will and Mae Wood. Mary Johnson provided information about Bachelor in an interview.

Phil Leggitt, former sheriff of Mineral County, took me on a tour of the Bachelor town site and identified several landmarks. I purchased a Bachelor saloon token (see fig. 60) from George Carpenter, a former Creede resident, and we had several subsequent telephone discussions about Bachelor. Doug Davlin graciously allowed me to include one of his mother's (Toni Davlin) sketches in the book (fig. 68).

Some of the photographs in this book were used in my earlier book, *Creede, Colorado History*. They were edited by William Schneider, the book's publisher.

Once again, it was a pleasure to work with Cheryl Carnahan as my copyeditor. She is knowledgeable, thorough, and has a magic touch that improved the quality and readability of the book. Her efforts are greatly appreciated. The publishers, Jan and P. David Smith, suggested several excellent changes that greatly improved the book.

Jane Morton graciously allowed me to use her poem that is included in the epilogue. I first heard her deliver this poem at a *Mining through Poetry, Stories, and Songs* event sponsored by the Creede Historical Society. Jane's daughter, Liz Morton Duckworth, alerted me to the *Saturday Evening Post* interview with Poker Alice and emailed me a copy.

Thanks to Phil Bethke (deceased), who gave me a copy of *Ancient Lake Creede*. He reviewed and made helpful comments on the text in Chapter 1.

Finally, thanks to my wife, Kay, who put up with me while I wrote the book and sometimes ignored her. She will be happy when the book is published and I am back to "normal" again. She also provided comments that helped improve the manuscript.

Chapter 1
Creede Caldera and Ancient Lake Creede

The history of mining and agriculture in the area of Bachelor and Creede, as well as the region's beauty, were determined by geological events that began approximately 35 million years ago. During this period the mountains surrounding Creede were formed and sculpted, and minerals such as silver, gold, zinc, and lead were deposited in fractured rock walls throughout the area. Tectonic activity at a later stage moved the Continental Divide to the west and created the fertile Upper Rio Grande Valley. It is important to understand these events because they created the Creede area that was discovered and developed by prospectors, homesteaders, mine owners, and business owners. Details, including maps and diagrams, can be found in *Ancient Lake Creede*, a compilation of scientific papers published by the Geological Society of America,[1] and *Creede, Colorado History*.[2]

The entire San Juan Mountain range is the result (after erosion) of volcanic and tectonic activity that took place in the southern Rocky Mountains between 5 million and 35 million years ago. During this period the Creede area was included in the Central San Juan caldera cluster, a subgroup of several volcanoes and associated calderas (fig. 4). A caldera is a large, roughly circular crater that is left after a volcanic explosion or collapse of a volcanic mountain. The Creede caldera is one of the best-preserved examples in the world.

The La Garita Caldera was by far the largest in the San Juan Cluster. After it collapsed about 27.5 million years ago, seven additional explosive eruptions and calderas formed within its boundaries over a period of about one million years.[3] Because molten rock (magma) from these volcanoes was very viscous, the eruptions were violent in nature—similar to the Mount St. Helens eruption in 1980—throwing huge clouds of rock and volcanic ash into the atmosphere. However, the San Juan eruptions dwarfed the Mount St. Helens event. The combined eruptions from the San Juan caldera cluster produced 2,000 cubic miles of ash flow—nearly 15,000 times that of Mount St. Helens.[4]

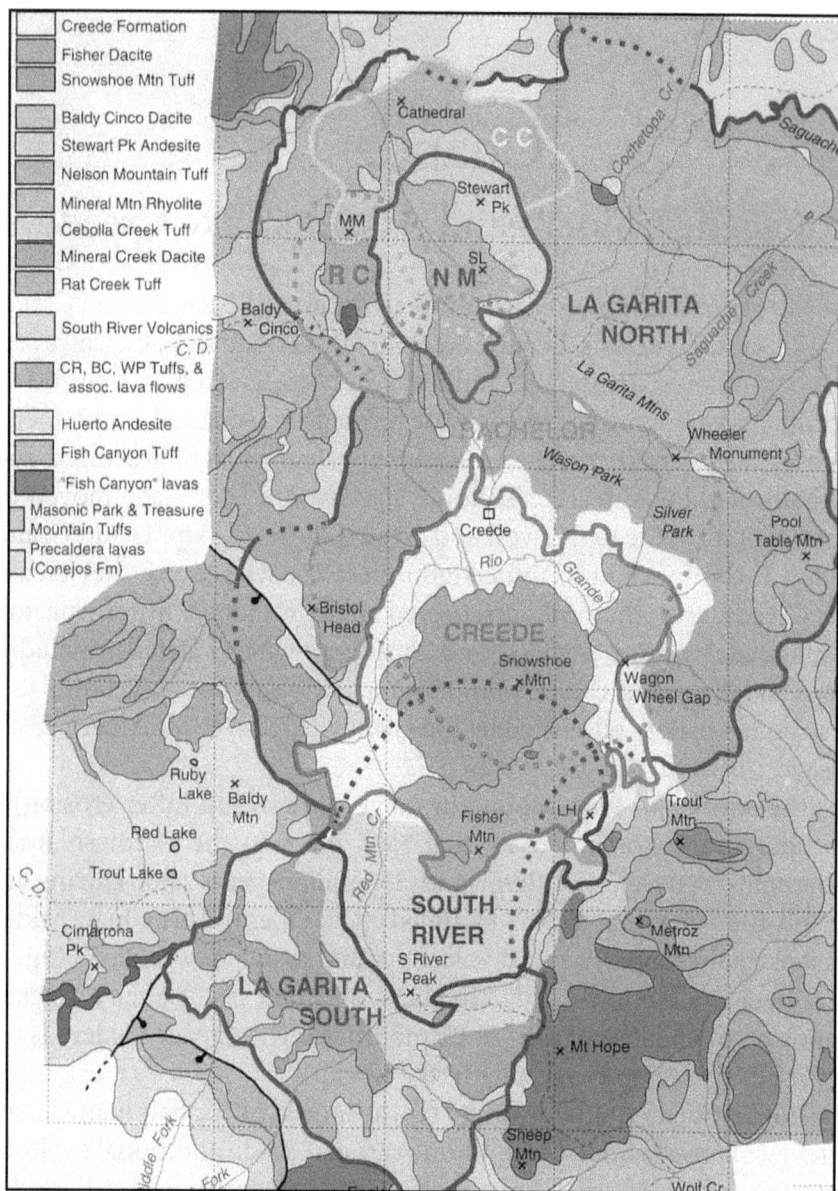

Fig. 4. Schematic of the Central San Juan Caldera Cluster. The figure shows the outline of the Creede Caldera (solid, roughly circular line extending from just south of Fisher Mountain at the bottom to just north of Creede at the top). The earlier-formed Bachelor Caldera (faint line between Wason Park and La Garita Mtns.) is seen surrounding the Creede Caldera to the north. The entire cluster is subsumed within the huge La Garita Caldera (dark and dotted line surrounding all calderas). From Bethke and Hay, eds., *Ancient Lake Creede*, frontispiece.

The Creede Caldera (outer solid and dotted line in fig. 5) formed approximately 27 million years ago and was the result of the last major eruption in the San Juan Cluster.[5] It overlapped a portion of the earlier Bachelor Caldera, which is partly responsible for the large cliffs north of Creede. This is important because mineral deposits in the famous Amethyst Vein were formed in these rock formations.

The Creede Caldera is approximately eight miles in diameter.[6] It has been studied extensively by geologists around the world because much of its outer edge is still visible today, particularly from space. The pronounced rock outcrops to the north of the Rio Grande are believed to be parts of the structural margins of the caldera.[7] The cliffs on the east side of Bristol Head Mountain mark the western wall of the Creede Caldera.[8]

A closed-basin lake, called Lake Creede (corresponding to the gray area in fig. 5), formed in the depression of the Creede Caldera. It varied in depth over time and covered approximately 200 km² (72 square miles).[9] Inflow of water was restricted by the western wall of the caldera (see later discussion), so the lake probably represented a steady state of inflow and evaporation. Plant fossils from this period can be found in the volcanic–ash–rich sediments in several places surrounding Creede. It is believed that the lake breached through Wagon Wheel Gap 21 million to 23 million years ago, draining through the Rio Grande Valley below this point.[10]

The final stage[11] in the formation of the Creede Caldera was a second magma upwelling that bowed the caldera floor upward to form what is known as a resurgent dome in the center of the caldera.[12] The dome rose a minimum of 1.5 km (about 5,000 feet) from the floor and, after erosion, created a nearly circular peak that is the present-day Snowshoe Mountain (fig. 6).[13] As the new mountain rose, Lake Creede deepened and formed an annular body of water circling Snowshoe Mountain from the lower Goose Creek drainage on the southeast to the lower parts of the Trout and Red Mountain Creek drainages on the southwest (see fig. 5).

The ancient lake was prevented from completely surrounding Snowshoe Mountain by volcanic domes along the structural margins of the caldera (e.g., Copper Mountain) to its south (fig. 5). At its peak, the surface of the lake was 10,800 feet above sea level. This is well above the current Creede and Bachelor topography because the area has undergone significant erosion in the millions of years since the lake was formed.[14]

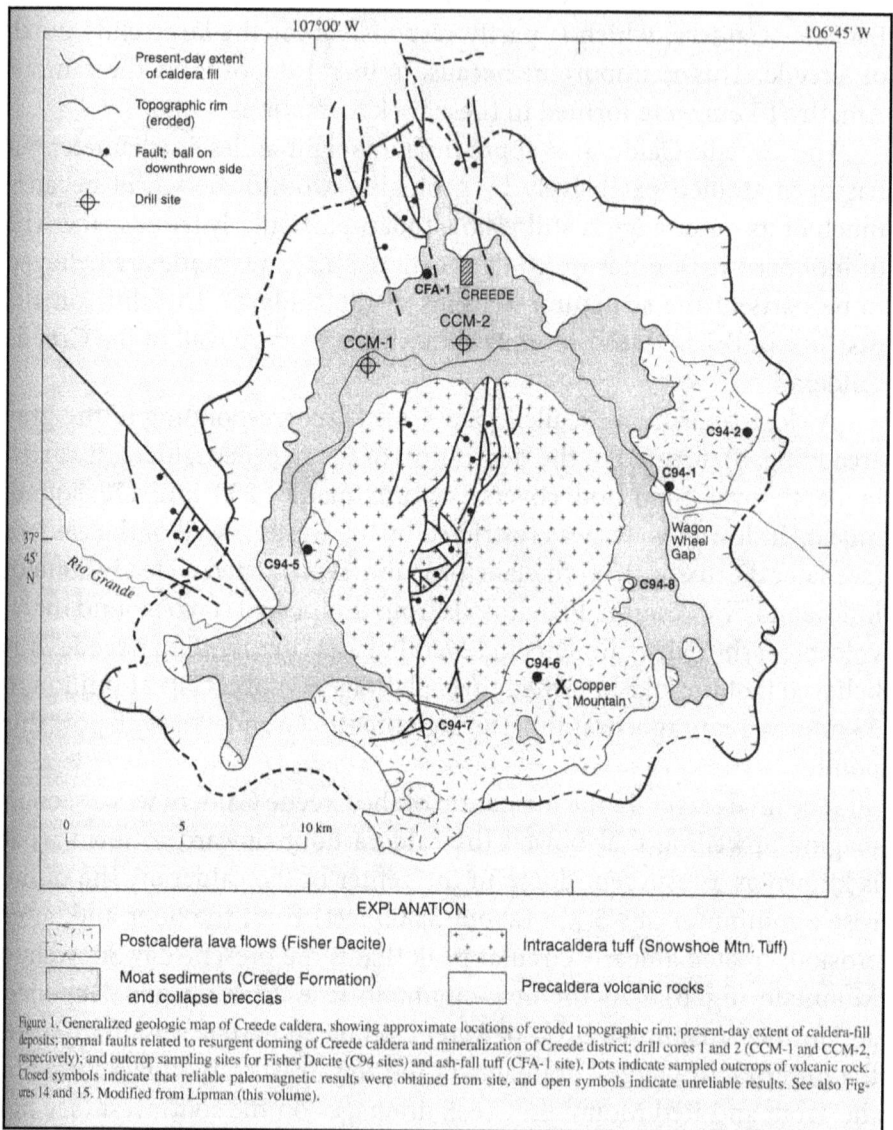

Figure 1. Generalized geologic map of Creede caldera, showing approximate locations of eroded topographic rim; present-day extent of caldera-fill deposits; normal faults related to resurgent doming of Creede caldera and mineralization of Creede district; drill cores 1 and 2 (CCM-1 and CCM-2, respectively); and outcrop sampling sites for Fisher Dacite (C94 sites) and ash-fall tuff (CFA-1 site). Dots indicate sampled outcrops of volcanic rock. Closed symbols indicate that reliable paleomagnetic results were obtained from site, and open symbols indicate unreliable results. See also Figures 14 and 15. Modified from Lipman (this volume).

Fig. 5. Schematic map of the Creede Caldera showing the original topographic rim. Creede is shown in the upper center. The dark gray is the Creede Formation, which corresponds very closely with the location of Lake Creede. As shown, it almost completely surrounds Snowshoe Mountain except on its southern margins.

Fig. 6. Snowshoe Mountain (noted by the arrow), the resurgent dome of the Creede Caldera, ca. 1930. The snowshoe formation, partially covered with snow, can be clearly seen on the mountain. Harbert Archives.

Extensive sedimentation into Lake Creede resulted from erosion, ash flows from active volcanoes, and sloughing from the sides of Snowshoe and other surrounding mountains. Sediments from the lake, which geologists refer to as the Creede Formation (gray area in fig. 5), are up to 1.26 km (about 4,100 feet) thick.[15] Over hundreds of thousands of years, water in the lake became a mineral-rich brine as a result of constant exposure to these sediments.

Lake Creede played an essential role in the history of Creede and Bachelor because it is believed to be a source of the fluids concentrated with silver, lead, zinc, and other minerals that were deposited in veins to the north and west of Creede. Philip Bethke and colleagues proposed a model for the deposit of minerals that involves circulation of subsurface water from Lake Creede and underground sources (fig. 7).[16]

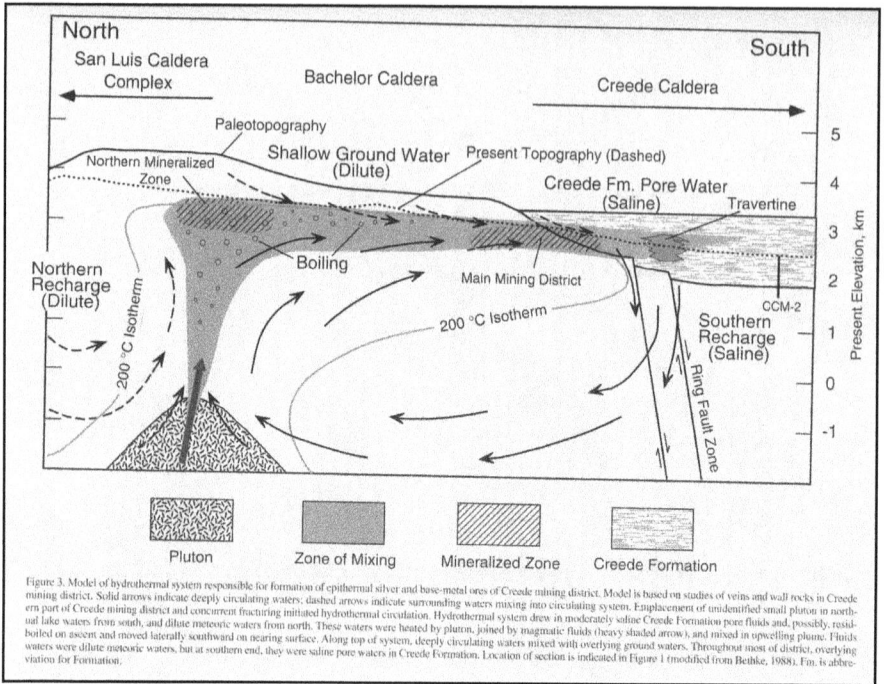

Figure 3. Model of hydrothermal system responsible for formation of epithermal silver and base-metal ores of Creede mining district. Solid arrows indicate deeply circulating waters; dashed arrows indicate surrounding waters mixing into circulating system. Emplacement of unidentified small pluton in northern part of Creede mining district and concurrent fracturing initiated hydrothermal circulation. Hydrothermal system drew in moderately saline Creede Formation pore fluids and, possibly, residual lake waters from south, and dilute meteoric waters from north. These waters were heated by pluton, joined by magmatic fluids (heavy shaded arrow), and mixed in upwelling plume. Fluids boiled on ascent and moved laterally southward on nearing surface. Along top of system, deeply circulating waters mixed with overlying ground waters. Throughout most of district, overlying waters were dilute meteoric waters, but at southern end, they were saline pore waters in Creede Formation. Location of section is indicated in Figure 1 (modified from Bethke, 1988). Fm. is abbreviation for Formation.

Fig. 7. Conceptual model of mineralization in the main part of the Creede Mining District. It shows the underground circulation of hot, mineral-rich brine that deposited minerals such as silver, zinc, lead, copper, and gold into cracks in the rock structures north and west of present-day Creede, including the Amethyst Vein. The model postulated that the source of energy to drive this circulation was heat from a magma chamber (pluton, lower left). Slight revision from Bethke and Hay, eds., *Ancient Lake Creede*, 305.

The authors postulate that a magma chamber formed north of Creede (under the Equity Mine area) approximately 25 million years ago, providing the heat and energy—and perhaps much of the mineral load—to create an underground convection current of mineral-laden brine through rocks north of Creede. These formations were created in the earlier Bachelor caldera. The hot water circulated through the fractures in the rock structures and, as it cooled, deposited minerals in them. Water returned to the lake, and the cycle repeated. This process occurred over hundreds of thousands of years and mineralized the Amethyst, Equity, Bulldog, and other veins north and west of Creede. These rich deposits remained undiscovered for another 25 million years until prospectors arrived in the area in the late nineteenth century A. D. This includes the rich Amethyst Vein that played such an important role in the formation of Bachelor.

Origins of the Upper Rio Grande Valley

The final chapter in the history of ancient Creede is the creation of the Upper Rio Grande Valley. Formation of the Creede caldera created a high rock wall to the west that diverted water to the north down Cebolla Creek into the Gunnison River drainage. This Ancestral Continental Divide lasted several million years until tectonic shifting 4 million to 5 million years ago uplifted the San Juan Mountains and tilted them to the east.[17] This shifted the Continental Divide about thirty miles west and allowed the Rio Grande to break through the southwest wall of the Creede caldera. The newly rejuvenated river then started flowing eastward and began the slow process of excavating the volcanic sediment from the valley. It is estimated that about seventy-five percent of the sediments from Lake Creede have been removed from the river valley.[18] Un-eroded portions of Lake Creede sediment are well exposed in the cliffs at Five-Mile Bridge near Antlers Ranch (Fig. 8).

Fig. 8. Cliffs at the Five-Mile-Bridge west of Creede, showing the exposed sediment and shale from Lake Creede. They are part of what is known as the Creede Formation. Harbert Archives, modified by William Schneider. More detailed discussions of early geologic events in the Creede area can be found, with additional diagrams, in references by Bethke, Harbert, and Huston in the bibliography.

Chapter 2
Beginnings
Early Inhabitants

The earliest inhabitants of the Upper Rio Grande Valley appear to have been from the Folsom culture, because they used the area as a summer camp and hunting ground approximately 10,600 years ago.[1] In 1977 Vincent Spero and Marilyn Martorano discovered a Folsom camping site on North Clear Creek that became known as the Black Mountain site because of its location on the slopes of Black Mountain. The first formal investigation of the site began in 1991 as a joint exploration by the Smithsonian Institution and the Rio Grande National Forest.[2] Excavations found several artifacts and confirmed that the campsite had been used by Folsom culture. It is believed they wintered at lower elevations and followed the game migration into this high-altitude camp. In fact, North Clear Creek appears to have been a well-traveled corridor between the Rio Grande and headwater trails along the Continental Divide.[3] How these early inhabitants reached North America is uncertain, but evidence suggests they might have crossed over a land bridge that existed between Siberia and Alaska about 14,000 BC.[4] They subsequently migrated south and eventually into Colorado.

Ute Indians inhabited the San Juans seasonally following the decline of the Anasazi culture in the fourteenth century.[5] The Utes first came in contact with Spanish explorers, from whom they acquired horses, in the early seventeenth century. Horses allowed them to travel longer distances during their annual migrations. They traveled in the spring of each year from lower elevations into the fertile river valleys of the San Juan Mountains.

The Mouache, Capote, and Tabeguache factions of the Ute culture likely crisscrossed the Continental Divide from the late seventeenth to mid-nineteenth centuries. This is inferred from the fact that early prospectors found numerous trails in the mountains when they arrived. Remains of Native American camps discovered at Wagon Wheel Gap and Phoenix Park were probably those of Ute summer camps.

Spanish explorers entered the Upper Rio Grande Valley beginning in the 1700s, but it is believed they did not explore the highest portions of the San Juans.[6] A map drawn by Don Bernando de Miera y Pacheco in 1778 shows the Upper Rio Grande and Animas River Valleys in reasonably accurate detail, suggesting they were familiar with the area—probably through contact with local Native Americans.[7] Hispanic settlement of the San Luis Valley began in the 1830s, stimulated by land grants made to secure Spanish claims to the region. The first permanent settlement was established in 1851 with the founding of the town of San Luis.

Individual fur trappers probably first entered the Upper Rio Grande Valley in the early 1800s. Trapping parties were known to be in the San Luis Valley in 1824 based on written communications.[8] Tom Boggs, brother-in-law of Kit Carson, settled in Wagon Wheel Gap in the summer of 1840.[9] He was probably the first permanent settler in the area surrounding Bachelor and Creede.

The first American exploration of the Upper Animas River Valley was in the summer of 1860. Charles Baker led a group of six men from the Leadville area down the Arkansas River and over Poncha Pass into the San Luis Valley. From there, they proceeded over Cochetopa Pass and worked their way up the Lake Fork of the Gunnison River and crossed over Cinnamon Pass into the Upper Animas River Valley. Despite promising early discoveries, the yield of gold was poor. While they identified rich outcroppings of silver, they could not exploit them at the time. They prospected in the valley that was later named Baker Park, then headed north out of the San Juan Mountains near the current town of Ouray. Baker sent reports of rich discoveries to Denver, which created intense interest in the area.[10]

In response to the positive reports, a second, larger group of prospectors left Leadville in December 1860. Although it was referred to as the "Baker Party" in some publications, Baker was not a member of the group. They arrived in the San Luis Valley in March 1861, then traveled south through Abiqui, New Mexico and north from there to the Upper Animas Valley. Again, they had little luck at mining the placer gold in the area. Instead, the rich ore was embedded in veins that would require expensive underground mining.[11]

No ore discoveries of note were made in the Animas Valley over the next several years because of the Civil War and the remoteness of the area. Baker returned to the valley for a short time in 1867, but was killed shortly thereafter in Arizona by Native Americans.[12] In 1869 a member of the original Baker party organized another unsuccessful prospecting party to revisit the area of the 1860–61 discoveries. Finally, when several members of the 1869 party returned in 1870, they discovered rich silver deposits near what later became Howardsville.[13] Readers interested in details of the early exploration of the Upper Animas Valley should refer to *The Story of Lake City, Colorado And Its Surrounding Areas* by P. David Smith (see Bibliography).

Stony Pass Route

Word of the silver strike in the Upper Animas River Valley spread quickly, and many prospectors found their way to the area the following year. However, access to the mines was difficult because the region was remote, rugged, and surrounded by steep, high mountains. One party came up to the headwaters of the Rio Grande Valley in 1869 and passed over the Continental Divide into the Animas Valley, thus establishing what became the Stony Pass route.

Stony Pass quickly developed as the primary route for transport of people and supplies to towns and mines in the Upper Animas Valley (fig. 9). The route was long and sometimes treacherous, but there were no good alternatives at the time. This was before there were any railroads in the valley. The Denver & Rio Grande Railroad did not reach Alamosa until 1878 and Del Norte until 1881.

Del Norte, located in the San Luis Valley, became the principal distribution center for freighters heading to the San Juans. By 1870, it had become the major commercial center in the west end of the valley, and was poised to become the supply center for people heading west.

The first road (more like a trail) over Stony Pass (12,588 feet above sea level) was built in 1872 by Major E. M. Hamilton, crossing the Continental Divide at its westernmost point in Colorado.[14] It was initially built to transport machinery for construction of the Little Giant Mill. Hamilton's crew proceeded across open country as much as possible, constructing wagon roads as necessary to traverse wooded or rocky areas. In the summer of 1872, approximately 150 people were prospecting in the Upper Animas Valley—triple the number from the previous year.[15] The number rapidly increased to over 1,000 by the mid-1870s. During the period 1872–82, thousands of people and many tons of supplies and mining machinery passed over the Stony Pass Route.[16]

Fig. 9. Approach to Stony Pass. The summit is approximately fifteen miles from this point up the Rio Grande Valley to the left. A narrow wagon road was all that existed in the 1880s. Before reaching this point, the road followed the river through the willow-lined valley that is now the Rio Grande Reservoir. Harbert Archives, modified by William Schneider.

Traffic was seasonal in the early years, with westward traffic in the spring and eastward migration to Del Norte or points east as soon as snow began to fall. Gradually, people settled in the area and spent the winters there.

A string of way stations and settlements was established in the Upper Rio Grande Valley to service freighters and travelers using the pass. Stage stations were built approximately every ten miles and usually had a horse barn, a blacksmith shop, and facilities for passengers to eat and freshen up. The town of San Juan City (near the present-day San Juan and Freemon Ranches) was established in February 1874 at the head of Antelope Park.[17] It was briefly the seat of Hinsdale County before it was moved to Lake City. A post office (opened in June 1874) and roadhouse, called the Texas Club, were also established near the site. San Juan City became a major storage and processing center where supplies were broken into smaller loads to be taken over the pass (fig. 10).

Fig. 10. Storage sheds at the old stage station at San Juan Ranch, ca. 1910. Joy Wills Nichols Collection, Creede Historical Society Archives.

Another major settlement on the route was Antelope Springs, located about ten miles southwest of Creede. It was also called Junction because the routes to Silverton and Lake City split at this point, with the Lake City route proceeding up Seepage Creek past Mirror Lake (now Santa Maria Reservoir) and the Silverton route continuing up the Rio Grande Valley to Stony Pass. At its peak, Antelope Springs included wayside services, a stage station, and several ranches. For travelers on the Stony Pass route, this was an important stop that included overnight accommodations. Antelope Springs was also known as "Aldens," named for George and Gustav Alden.[18] The Alden brothers established a post office at Antelope Springs in May 1876 and built a hotel and store building to enhance its attractiveness to travelers. The Alden properties were sold to Jackson Soward in the spring of 1879. Soward and his sons developed a sawmill in Antelope Park, assumed post office responsibilities, and expanded wayside services.[19] As the Stony Pass route declined, the Soward family became some of the first homesteaders in the Upper Rio Grande.

The Barlow and Sanderson Stage Line, which ran from Del Norte to Lake City, used Antelope Springs as a station on its route.[20] The stage line also used the Rio Grande Station, located a few miles south of Creede near the east end of the Deep Creek cutoff, close to the present-day Wason Ranch. Of course, there was no town of Creede at this time; it didn't come into existence until 1890.

A hotel was opened at Wagon Wheel Gap in 1876 to service people hauling goods from Del Norte to the San Juan mines.[21] In 1881 William J. Hare homesteaded 160 acres at the junction of Goose Creek and the Rio Grande and established a ranch that also became a stage station on the Del Norte to Silverton and Lake City road.[22]

The D&RG Railroad extended its tracks to Wagon Wheel Gap in 1883, after the demise of the Stony Pass route. The extension was sponsored by General William Jackson Palmer, founder of the railroad, to ferry tourists to the hot springs located up Goose Creek. He took an interest in the area because he liked to entertain guests, and the springs were believed to treat arthritis symptoms. The springs included a large bathhouse and a hotel. At one time Palmer owned much of the land that eventually became part of the current 4UR guest ranch.[23]

When the D&RG Railroad reached Silverton in 1882, traffic over Stony Pass slowed abruptly and the way stations to support it went into decline. Many people abandoned the area, but a hardy few stayed on to homestead or prospect. Some of the abandoned way stations later became nuclei for larger ranches. However, for Creede and Bachelor the story was just beginning. The mining activity in the western San Juans combined with dramatically increased traffic had opened the Upper Rio Grande Valley to both mining and agricultural possibilities. The stage was set for the discovery of one of Colorado's great silver mining camps.

Discovery of the Amethyst Vein

The impetus for the formation of the town of Bachelor was the discovery of claims that were developed into mines along what would be called the Amethyst Vein, located on Bachelor Mountain in the Sunnyside Mining District. These mines, just a short walk from the town of Bachelor, became the most productive mines in the Creede Mining District and fueled the mining boom that ensued. The story of the discovery of silver in the Upper Rio Grande Valley was detailed by W. H. Emmons and E. S. Larsen in 1923.

In the 1870s and early 1880s the upper part of the of the Rio Grande valley was a route of transportation between Wagon Wheel Gap and the flourishing camps near Silverton and Lake City. This route passed very near the present site of Creede and nearer still to Sunnyside, a small camp about 2 miles west of Creede. Some prospectors halted at Sunnyside to investigate the steep mountain slopes along the valley, and finding encouraging indications, located several claims. J. C. MacKenzie and H. M. Bennett located the Alpha claim at Sunnyside April 24, 1883, and with James A. Wilson located the Bachelor claim, near the present site of Creede, July 1, 1884. Some prospecting was done in the middle eighties, principally at Sunnyside, and futile attempts were made to work the ores in arrastres (primitive mills for grinding and pulverizing rock). As early as 1885, Charles F. Nelson prospected the site of Creede and, getting only small returns, went to Sunnyside. Richard and J. N. H. Irwin bought the Alpha claim in 1885 and located other claims nearby. There is no record of any new discoveries from 1886 until August, 1889, when N. C. Creede, E. R. Naylor, and G. L. Smith located the Holy Moses claim on Campbell Mountain. The next summer Creede located the Ethel and C. F. Nelson located the Solomon claim. The mining district that was formed was called the King Solomon district; it is east of and nearly continuous with the Sunnyside district.

The promising assays that were obtained from the Holy Moses claim, which was christened from the exclamation Creede made when he first beheld the outcrop, led to the rapid development of the district. When it became generally known that Creede had sold an interest in the Holy Moses Mine to D. H. Moffat and associates, of Denver, prospecting was renewed with great vigor.

In June, 1891, Theodore Renniger and Julius Haase, grubstaked by Ralph Granger and Eric Buddenbock, two butchers of Del Norte, set out to prospect the region of Creede. It is said that a search for their strayed burros led Renniger to the outcrop on the Last Chance claim, which was located in August, 1891. Creede, who was then engaged in developing his claims on East Willow Creek, visited the site of the discovery and traced the outcrop for some distance. Impressed with the surface indications Creede prevailed on Renniger to define his claim (the Last Chance), and then located next to it the Amethyst claim, in the names of D. H. Moffat, L. E. Campbell, and himself. Haase sold his interest in the Last Chance claim to his partners for $10,000, and in November, 1891, Renniger and Buddenbock sold their thirds to investors in Leadville and Denver. Granger, one of the original locaters, was offered $100,000 for his third interest but did not sell.

In April, 1891, J. C. MacKenzie and W. V. Gilliard located the Commodore, and in August of that year George K. Smith and S. D. Coffin located the New York as the southerly extension of the Last Chance. This was on the Amethyst or "Big" Vein, upon which the Bachelor claim has been located six years before, but the two locations were nearly three-quarters of a mile apart. Within a few months the Amethyst Vein was pegged for a distance of nearly two miles along its strike. Mammoth, Campbell, and MacKenzie Mountains each received a due share of attention from numerous prospectors.[24]

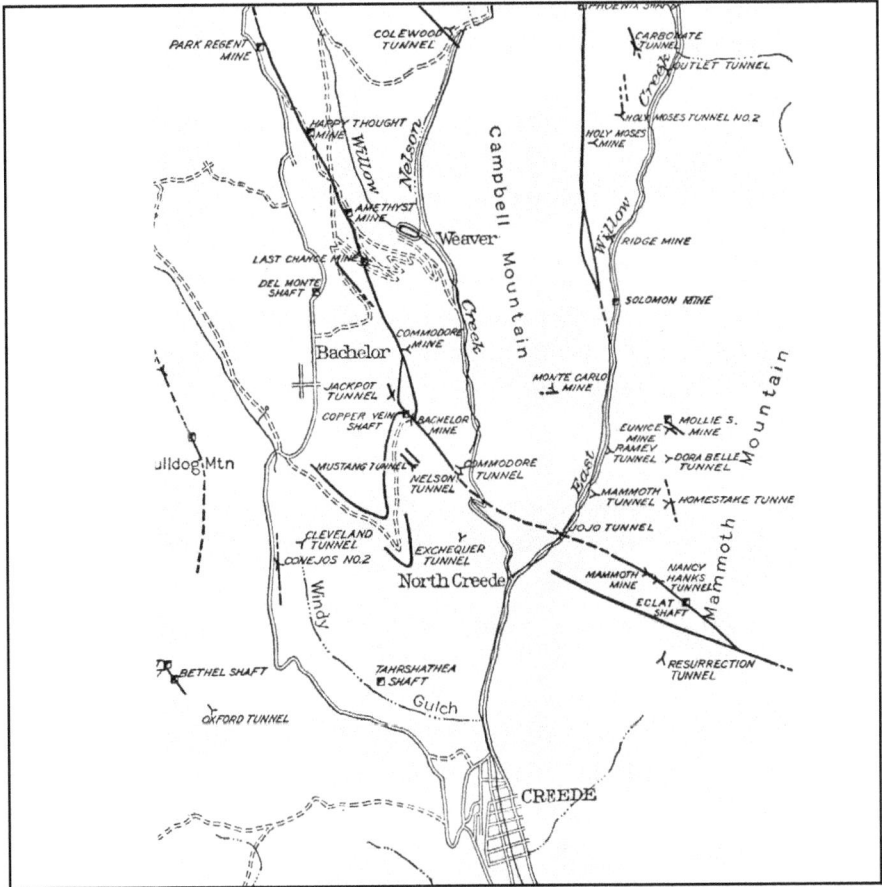

Fig. 11. Map of mines in the Creede–Bachelor area. The Amethyst Vein is the solid line that begins at the Commodore Tunnel in the center of the map and progresses in a northwesterly direction toward the Park Regent Mine. The town of Bachelor is northwest of Creede. Geological Survey Bulletin 811-B, plate 26. Harbert Archives.

The rapid succession of discoveries in 1891 on the Amethyst Vein fueled the boom that led to the rapid growth of Creede and the founding of Bachelor. The map in figure 11 shows the mines on the Amethyst Vein and their locations relative to the towns of Bachelor and Creede. In retrospect, the long hiatus between the discovery of the Bachelor claim in 1884 and those of the Last Chance and Amethyst in 1891 is surprising.

Three years earlier, MacKenzie had located several claims on a large vein of quartz about 150 feet above the Amethyst lode. These claims were abandoned until the fall of 1891, when the Del Monte location covered them. Charles W. Henderson provides a clue as to why MacKenzie missed the outcrops on the Amethyst Vein: "Debris from the upper part of the mountain had so deeply covered the Amethyst lode as to prevent its recognition earlier."[25] This suggests that Renniger and Haase were either very diligent or lucky in their efforts that led to the discovery of the Last Chance claim. Huston states that they were "just plain lucky."[26] Sam Coffin, one of the co-founders of Bachelor, was another fortunate person because he had the rare luck to be on the ground floor when the Last Chance and Amethyst claims were discovered. As mentioned, he discovered the New York claim on the south boundary of the Last Chance claim. The Last Chance owners amended their survey to include the New York claim and started litigation to clear the title of their claim. Eventually, a compromise was reached to resolve the dispute between the two claims, resulting in the formation of the Consolidated New York Company (fig. 12).[27]

Following discovery of claims along the Amethyst Vein, the next phase was development of the mines. This led to a dramatic increase in the number of miners, capital investment, and extraction of ore from the mines. It was during this peak activity phase that the town of Bachelor was formed. Details of the development of the mines will not be discussed here, but an excellent discussion can be found in *A Silver Camp Called Creede* by Richard C. Huston.[28]

Fig. 12. View of the New York–Chance Mine showing teamsters and their ore wagons. This was one of the closest mines to Bachelor. The photograph was taken in 1893 by Brooks and Drake of Creede. Harbert Archives.

Chapter 3
Birth of a New Mining Town
John MacKenzie

John MacKenzie can rightly be considered the "Father of Bachelor."[1] As mentioned, he came to southwestern Colorado in 1876 and set up camp on Rat Creek, in what was later called Sunnyside. He must have crossed back and forth many times through the meadow that later became Bachelor. He built a cabin there and recognized it as a desirable place to build a town. MacKenzie was one of the original pioneers who was respected and held in the highest esteem, and his role in founding Bachelor was recognized by those who followed him. He was an old settler long before Creede came into existence (fig. 13).[2]

In its first issue, Bachelor's newspaper, *Teller Topics*, acknowledged MacKenzie's importance to the existence of the town: "The upbuilding of Bachelor pays tribute to the wisdom and acumen of J. C. McKenzie, who fourteen years ago laid the foundation of the city by building his cabin on the hill near to the prospect and exploiting the spring in the park where the camp stands today."[3]

MacKenzie's role in the establishment of Bachelor and, in fact, the entire Creede Camp has frequently been understated, especially in reference to the greater publicity accorded to Nicholas C. Creede. Richard S. Irwin, another early prospector and pioneer, paid tribute to MacKenzie in a January 1893 letter to the *Creede Candle*: "Mr. MacKenzie, an old and successful Georgetown and Geneva, Colo., miner deserves the credit of keeping this camp alive for ten years, and until Mr. Creede came over the range . . . Yours, etc., R. S. Irwin".[4]

A more detailed account of MacKenzie's contributions is found in *Mines and Mining Men in Colorado*:

Fig. 13. John MacKenzie, the "Father of Bachelor."
Creede Historical Society Archives

After spending some time in Georgetown he started for the scene of the San Juan excitement, intending to go to Silverton, but incidents brought him into Sunnyside, one of the sections of Creede camp. Finding the country to his liking he camped in Rat Creek gulch and prospected about the surrounding hills, locating the Bachelor, Commodore, Spar, Copper, Mustang and Del Monte on Bachelor mountain, and the Combination, Wisconsin Boy, Maid of Sunnyside and Alpha on Mackenzie mountain. The Bachelor is proving to be the richest producer in the camp. The Alpha is also a shipper of good proportions, and the other properties are in a fair way of becoming big dividend payers for those who purchased from Mr. Mackenzie.[5]

MacKenzie's death in October 1894 was reported in the *Creede Candle*:

John C. McKenzie, well known to nearly all the people of Creede camp and the mining men of the west, died at Del Norte at 1 o'clock a. m. of Saturday, October 13, of dropsy from which he had been a sufferer for many months.

The article continued:

He left no will so far as can be learned and an administrator will have to be appointed to attend to the estate. Was unmarried and the only known relative is a brother in Halifax. . . The death of Mr. McKenzie removes one from the ranks of the old pioneers who was respected by all and held in the highest esteem as a man, a citizen and a friend.[6]

Origins of a Mining Town.

Prior to the major discoveries on the Amethyst Vein, the active mines could be easily reached from nearby communities by foot, on horseback, or by riding on wagons. However, with the discoveries of the mines on the steep slopes of Bachelor Mountain, access to these properties became more difficult. Miners had to walk or ride up the steep and narrow West Willow Creek Canyon, then climb up to the mines. Alternately, they could take a path/road from west of Creede up Bachelor Mountain. In the winter, the trip was even more difficult and sometimes impossible. The need for a town in closer proximity to the Last Chance, Amethyst, Happy Thought, Bachelor, Park Regent, and other mines on the Amethyst Vein was critical.

There are slightly different versions of how Bachelor was founded and who first settled there. It was an ideal place for development because it was an open meadow with at least two freshwater springs. The fact that John MacKenzie had already built a cabin on Bachelor Mountain was mentioned in the *Teller Topics* article cited in the introduction to this chapter.

Mineral County did not exist—it was not founded until March 1893—when mining activity began in the area. In fact, the Creede Mining District was at the intersection of Saguache, Rio Grande, and Hinsdale Counties (fig. 14). This created special problems when a miner decided to make a claim or an owner wanted to register a mine because of the overlapping jurisdictions. To avoid possible lapses, people sometimes filed in all three counties when the jurisdiction was in doubt. The Bachelor town site was clearly located within the boundaries of Hinsdale County, so all legal filings were made in Lake City.

As mentioned, living in Creede was impractical for miners working the mines on Bachelor Mountain, which was nearly two miles away up the steep West Willow Creek Canyon. Those who braved this route still faced a daunting, several-hundred-foot climb above the canyon floor. The communities of Weaver and Stumptown grew up spontaneously nearby, but neither had the open space needed to house a large population. The long, sloping meadow just a short walk over a gentle hill from the mines became a logical alternative.

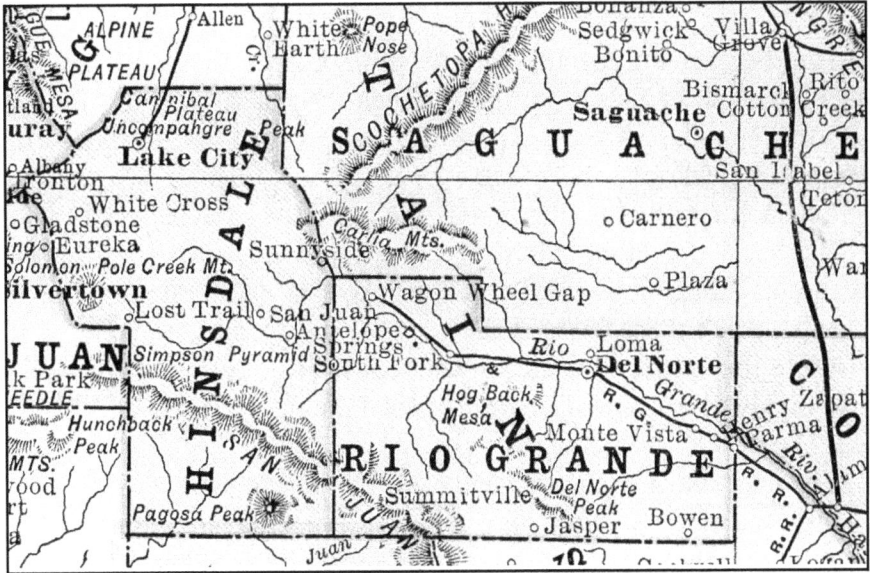

Fig. 14. Intersection of Saguache, Hinsdale, and Rio Grande Counties before Bachelor and Creede came into existence. When formed in 1891–92, Bachelor was in Hinsdale County near Sunnyside. Map by Mast, Crowell & Kirkpatrick, ca. 1890. Harbert Archives.

A detailed description of Bachelor's founding was published in the September 11, 1892, issue of the *Colorado Sun*:

Bachelor, Colorado., Sept. 10.—Bachelor is in Hinsdale county, on the east slope of the main range of the Rocky mountains, and, more directly speaking, on the southwest side of Bachelor mountain, in a park of some two miles in length. It is two miles and a half from Creede or Jimtown, on the road built by the Last Chance Mine.

The camp first started by being the only park near and on the [Last] Chance road, and on Sept. 6, 1891, C. L. Calvin, with his wife and family, settled here and built a dwelling and boarding house, near two springs of water. Soon after the Last Chance struck pay mineral and put on a force of miners, and then they built boarding and bunk houses near the springs.

Next came the livery and feed stables by C. Pierson. Following this came the inevitable saloon by a Mr. Carter, and from this on the town grew. Stores and saloons opened with a rapidity only found in the West. The population increased very rapidly, so rapidly that a short time ago it was decided to hold a special election with a view of incorporation.[7]

Another version credits Sam Coffin, prospector friend of John MacKenzie, as the first to settle in Bachelor (fig. 15): "S. D. Coffin was one of the pioneers of this new town, having built the first house in December, 1890."[8]

Yet another version states that Coffin was the second person to build in Bachelor: "Mr. Coffin built the second house in Bachelor City, where he and his family now reside."[9] Regardless, it is clear that he is widely considered one of the town's earliest founders.

Once the first few buildings were constructed, the town began to grow rapidly. The *Creede Candle* described the rush to build on the site:

The latest town site excitement is in the park on Bachelor Hill, around the Last Chance boarding house. Yesterday there was a stake raid for lots and it is proposed to plat an 80 acre government town site to include the territory of the Last Chance mill site, portions of the Quakenasp and Greenback claims and some vacant ground. Two saloons and a female seminary are already in operation and other business houses expected. It is to be called Bachelor.[10]

Fig. 15. Sam Coffin, one of Bachelor's founders and an early settler in the town.
Creede Historical Society Archives.

The listing of a "female seminary" as one of the establishments already in operation is amusing because there is no record that such an establishment existed in Bachelor. One is left to wonder if this was a euphemism for a brothel. If so, the editors of the *Creede Candle* displayed a sense of humor.

Driven by the need to get ahead of the excitement and provide direction for the pending housing boom, Sam D. Coffin and John MacKenzie joined in the project of building an orderly town on the site.

Sam D. Coffin, satisfied with the mineral outlook of Bachelor, joined McKenzie in the project of building a town in the clouds. In January, 1892, Mr. Coffin, G. C. Martendale, and Newt. J. Thatcher had a survey made for a townsite which included eighty acres, platted into twenty-four blocks, with three main streets and nine cross streets.[11]

In the formal plat filed in Hinsdale County several months later (see discussion below), the three streets running south to north were labeled Main in the center, Aspen to the east, and Park to the west. A fourth street to the extreme west was unnamed. The cross streets, running east and west, were North Avenue to the extreme north and First through Eighth Avenues counting to the south.

C. C. Davis, a Leadville newspaperman, braved the winter and snow in March 1892 to visit the bustling new town. His description in Leadville's *Herald Democrat* captured the excitement at the time. It is also evident from his writing that many of Bachelor's new residents had migrated from Leadville to pursue the latest mining boom:

On approaching the summit of Bachelor mountain I was somewhat astonished to find myself in the midst of a wilderness of new roofs and a bustling community; but before me was a well-laid-out village of one hundred houses, while others were going up on all sides. This is Bachelor City, intended to be the home of the working miners, and, although but a few days old, presented a robust and inspiring spectacle. The recorder of the village is Tom Vincent, an old Leadville boy, and he informed me he had that day recorded one hundred and seventy-five sales of lots. A little further on, the extensive saw mill of Alderman A. S. Crawford, of Leadville, came in view, and I began to feel somewhat at home. Half a mile beyond, on the northern slope of a very steep mountain, the shaft houses of half a dozen properties are visible. Chief among these are the Last Chance and the Amethyst, while further east lies the Holy Moses.[12]

The new town began to grow rapidly as more and more miners, businessmen, and their families streamed to the city on the hill (fig. 16). Without a newspaper in Bachelor to capture the excitement, reporting fell to the *Creede Candle*. In its April 8, 1892, issue, it published this story:

Bachelor City is one of the liveliest of the many towns in the camp. Many new buildings are going up and the merchants are having a good trade. There are eight stores, about a dozen saloons, several assay offices, boarding houses, hotels and restaurants in operation and more going up. Some large and substantial two-story buildings are in course of erection, and the town has all the appearances of permanency. The order for a resurvey of the township and some technacalities [sic] in the townsite papers have delayed getting titles to lots, and the fact that the town is on mineral ground will operate against, but there is no doubt but that Bachelor will be one of the best suburbs of Creede.[13]

The potential for a major fight over land rights to the town site was evident because the new plat overlaid several mining claims that already existed in the Sunnyside Mining District. In a map drawn in June 1892 by E. S. Rice, parts or all of the Quaking Asp, White, Equitable, Greenback, and Little Penn claims clearly appear to be within the town limits (fig. 17). While the town boundaries are not delineated on the map, the Legal Tender, Maid of Erin, B. B. Simmons, and Big Ingun claims may have also overlapped the town site.

A notice in the April 22, 1892, issue of the *Creede Candle* highlighted impending legal problems: "All parties are warned against occupying or building upon that portion of the townsite of Bachelor in conflict with the White and Equitable mining claims. Rocker & Spurgeon, Attorneys for Owners."[14] The same notice regarding the White and Equitable mining claims appeared in several issues of the *Candle* in the spring of 1892. In the June 24 issue, a similar notice appeared warning against building on the Quaking Asp claim: "All parties are warned against occupying or building upon the Quaking Asp mining claim, a part of which is covered by the townsite of Bachelor. T. E. McClelland, Attorney for Owner."[15]

Despite all the ominous signs, it appears that no successful legal cases were brought against the town of Bachelor. There could be several reasons for this outcome, but the most likely is that the mining claims proved worthless.

Fig. 16. Panorama of Bachelor, probably taken from Bulldog Mountain, in its early days (likely spring 1892). The trees have been cleared extensively to support mining activity on the Amethyst Vein located over the brow of the hill behind the town. Gardanier Collection, Creede Historical Society Archives.

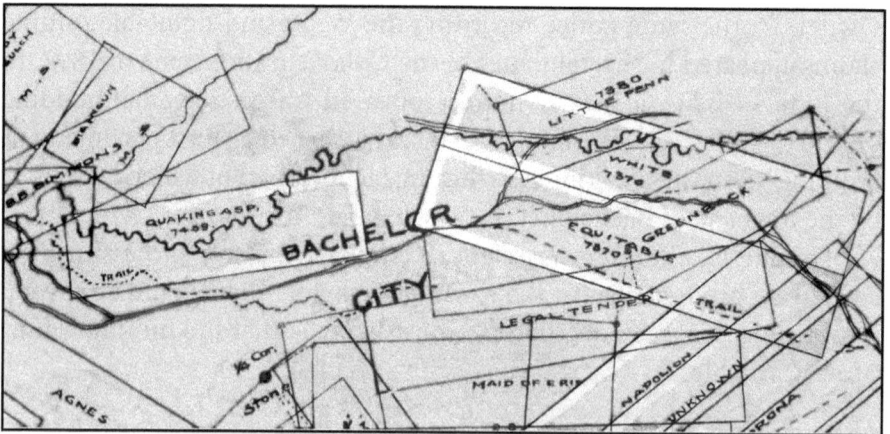

Fig. 17. Portion of a cyanotype (blueprint) map prepared by E. S. Rice of South Creede in June 1892. Although the map is in poor condition, the overlap of several mining claims and the town of Bachelor City is apparent. Creede Historical Society Library.

Bachelor Gets a Post Office

As the town continued to grow Bachelor needed its own post office. Normally, this would be a straightforward process, but it was complicated by the prior existence of a post office in California also named Bachelor. Located in Lake County, it had been established in 1882, ten years earlier. Because the typical abbreviations for Colorado and California at the time were "Col" and "Cal," confusion was almost certain, particularly with handwritten addresses. Early mention of this problem appeared in the April 22, 1892 issue of the *Creede Candle*: "Bachelor City will have a postoffice in a few days. All has been decided upon but the name, and that had to be changed because of their [*sic*] being a Bachelor in California."[16]

In the end, the issue was resolved successfully and the post office was named Teller (fig. 18) for Henry M. Teller, one of the Centennial State's most prominent statesmen. Teller, who established a law practice in Central City in 1861, was elected one of the first two senators when Colorado became a state in 1876. After briefly serving as Secretary of the Interior under President Chester Arthur, he was reelected two more times.[17]

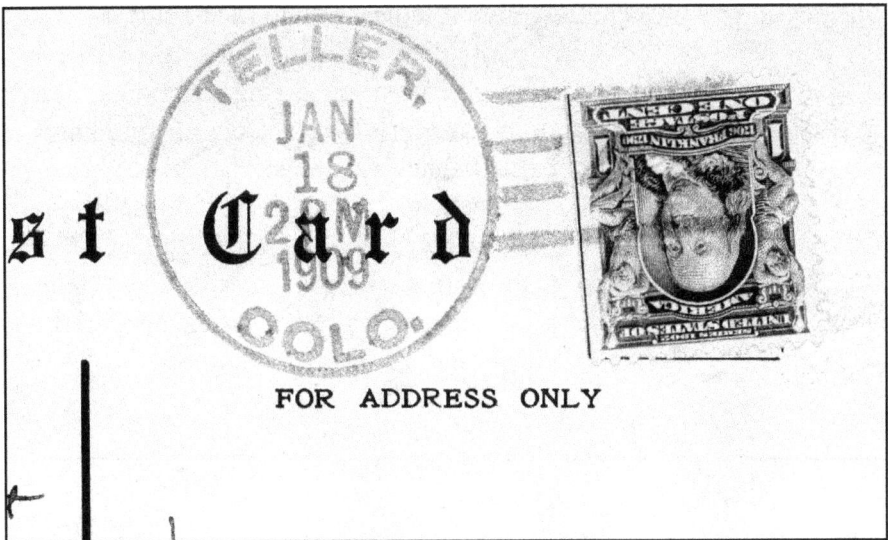

Fig. 18. Teller postmark on a postcard mailed January 18, 1909, from Bachelor to Monte Vista, Colorado. Because the post office was open for only twenty years, this is a rare postmark. Harbert Archives.

An article in the May 13, 1892, issue of the *Creede Candle* celebrated the new post office in Bachelor: "The postoffice at Bachelor City has been named Teller and John Gould appointed postmaster. Hurrah for Teller!"[18]

In its first issue, *Teller Topics*, Bachelor's first newspaper, praised the selection:

> It is a well-deserved tribute to one of the cleanest and best servants Colorado ever had, paid in naming the postoffice after Senator Henry M. Teller. The men who suggested the name honored themselves and their town. It is peculiarly fitting that this testimonial should come from the greatest silver camp in Colorado to the bravest and truest silver advocate in the United States senate. Senator Teller will need nothing to perpetuate his name to the people of Colorado other than his own great work, but Creede camp needs Senator Teller, and this tribute is but a slight return to him for his fight for the camp's great industry.[19]

The post office got off to a good start, as reported in the June 24, 1892, issue of *Creede Candle*: "Teller post office is now established at Bachelor and mails are arriving and departing daily. John Gould fills the position of postmaster with evident satisfaction to the public."[20]

The news of the Teller post office was apparently slow to reach people in the county seat of Lake City. An imploring question appeared in the June 2, 1892 issue of the *Lake City Times*: "Does anyone know of a postoffice in this county named Teller? Postmaster Steinbeck has been receiving mail matter the past week directed to 'Teller, Hinsdale county, Colorado.'"[21] This is perhaps excusable given the long distance between the two towns, but it is an indication of Lake City's lack of attention to the new town and its post office. Ironically, the article appeared one week after Bachelor was incorporated in Hinsdale County.

Chapter 4

Incorporation

The next major item of business for the new town was incorporation, mentioned briefly in the April 22, 1892, issue of *Creede Candle*: "The proving up of Bachelor City will now be done in Hinsdale County."[1]

At the time of its formation, before Mineral County came into existence, various parts of the Creede Mining Camp were in Hinsdale, Saguache, and Rio Grande Counties. The Sunnyside Mining District, where Bachelor was located, was on the boundary between Hinsdale and Saguache Counties. Because its location fell slightly on the Hinsdale County side, the town initiated the process with the Hinsdale County Court in Lake City.

An original set of three documents requesting incorporation was filed on April 13, 1892. They are presented here in the order in which they appeared in the filing package.[2] The first was a cover letter prepared by Spencer & Watson, attorneys for the petitioners (fig. 19).

Cover Letter for Incorporation Documents

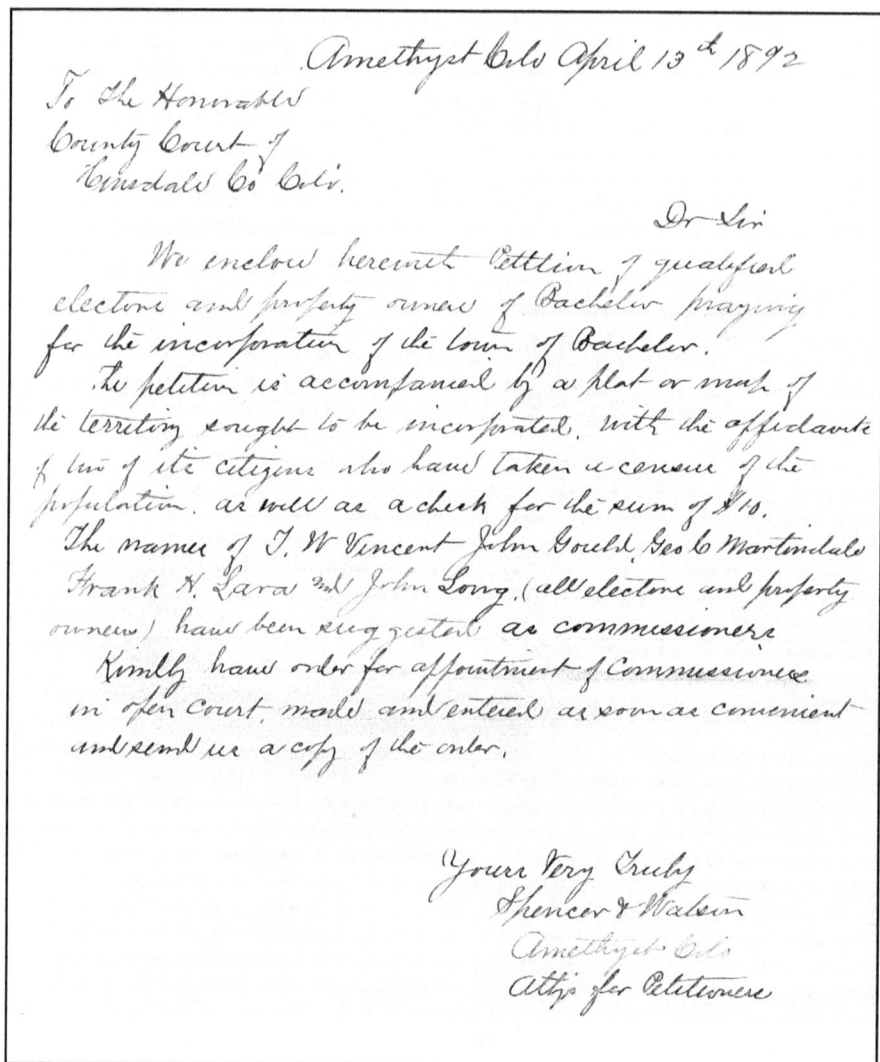

Amethyst Colo April 13th 1892

To The Honorable
County Court of
Hinsdale Co. Colo.

Dr Sir

We enclose herewith Petition of qualified electors and property owners of Bachelor praying for the incorporation of the town of Bachelor.

The petition is accompanied by a plat or map of the territory sought to be incorporated, with the affidavits of two of its citizens who have taken a census of the population, as well as a check for the sum of $10.

The names of J. W Vincent John Gould Geo C Martindale Frank A. Lara and John Long, (all electors and property owners) have been suggested as commissioners.

Kindly have order for appointment of Commissioners in open court, made and entered as soon as convenient and send us a copy of the order.

Yours Very Truly
Spencer & Watson
Amethyst Colo
Attys for Petitioners

Fig. 19. Cover letter, dated April 13, 1892, for documents petitioning the incorporation of the town of Bachelor. Courtesy, Hinsdale County Museum, Lake City, CO.

Certification of Canvass of Town of Bachelor

The second document (fig. 20) was a affidavit certifying that Ed O'Kelley and Joseph Either had canvassed the proposed town to get an accurate count of the number of residents. This was a necessary first step because the legal requirements were different depending on the size of the community requesting incorporation. It was not important for the two canvassers to come up with the same exact count, which they did not. The affidavit is reproduced here.

State of Colorado
                        state seal
County of Hinsdale

                        In the County Court
                        in and for said County

    Ed O'Kelly and Joseph Either being first duly sworn on oath each for himself and not one for the other says [*sic*] that they have made an accurate and complete canvass of the inhabitants of the territory embraced within the limits of the town of Bachelor in said county as shown by the map and plat hereto annexed:
    That applicant Joseph Either canvassed the eastern portion of said town and found within the district which he canvassed a population of Three hundred and forty seven persons and no more.
    That applicant Ed O'Kelly canvassed the western portion of said town and found within the district which he canvassed a population of Three hundred and sixty two persons and no more.
    Applicants further make oath that they are citizens and property owners within the territory embraced within the limits of said town, and believe that the population as stated above, the result of a personal canvass, comprizes [*sic*] the full number of inhabitants of said territory on the 13th day of April A D 1892.

                        Ed O'Kelley (signature)
                        Joseph Either (signature)

Fig. 20. Record of the canvass of Bachelor that was part of the request for incorporation. Courtesy, Hinsdale County Museum, Lake City, CO.

O'Kelley[3] and Either signed the document in their own handwriting. The former was deputy sheriff (or marshal) in the unincorporated town of Bachelor at the time. His name is spelled O'Kelly in the body of the document, but he signed his name "Ed O'Kelley." It isn't clear from the letter whether the counts of O'Kelley and Either were independent counts of the entire town or of just the western and eastern portions, respectively. Thus the population of the town at the time of canvass was either approximately 350 or 700, depending on the interpretation.

The third part of the April 13 application was a legal description of the town followed by the signatures of the thirty-three petitioners. The first section of this document, including the metes and bounds of the town, is not included for the sake of brevity. However, part of the text and the signatures of the thirty-three men are shown in Figure 21. These people, as well as those who participated in the vote for incorporation (see below), are referred to as founders in later citations in the book.

Signatures of Petitioners for Incorporation of Bachelor

To the County Court within and for the County of Hinsdale, State of Colorado: . . .

Therefore, We, the undersigned, duly qualified electors and land owners, residing and owning land within the limits of the territory to be embraced in the proposed incorporated town, hereby respectfully petition the Honorable County Court of Hinsdale County, in the State of Colorado, to forthwith appoint five Commissioners, as by law required, to call an election of all the qualified electors residing within the territory embraced within the limits described herein, and in the map or plat hereto annexed, to vote upon the question of incorporation of said territory as an incorporated town and to do and perform, as such Commissioners, all other acts required by statute in such case made and provided, to incorporate such town and of holding the first election of officers therefor, in case the said territory shall be incorporated. . .

And your petitioners further and respectfully represent that this petition, map and plat is [sic] accompanied with the proofs required by law, of the number of inhabitants embraced in said limits, and your petitioners say that said territory contains a population less than two thousand, and must therefore, under and by virtue of the statutes in such case be made and provided, be organized as an incorporated town.

And now your petitioners further and respectfully represent that the name proposed for such incorporated town is "Bachelor."

And for all of which your petitioners will ever pray.

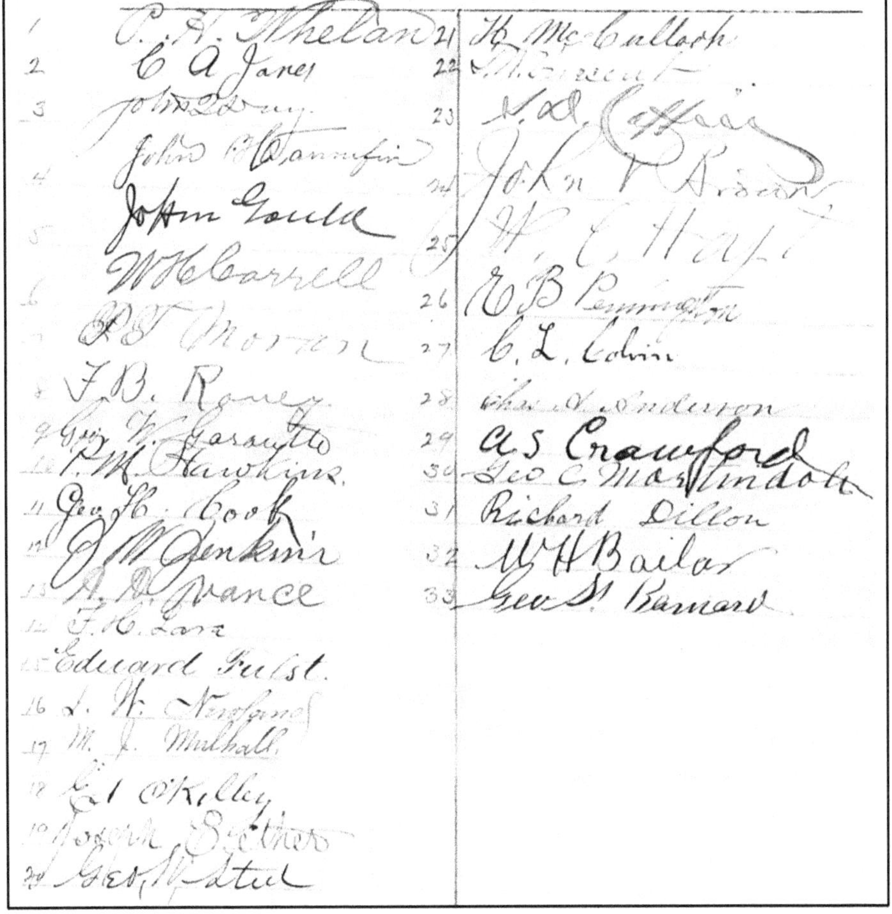

Fig. 21. Signatures of residents who signed the petition for Bachelor's incorporation. S. D. Coffin (signature 23) paired with John MacKenzie to found the town. Gould (5), Jenkins (12), Newland (16), and Crawford (29) were elected to the first Board of Trustees (see below). Vincent (22) became town recorder and Edward Fulst (15) was later elected mayor. Signature 18 is that of Ed O'Kelley, who murdered Bob Ford in Creede two months later. His accomplice was Joseph Either (19). Courtesy, Hinsdale County Museum, Lake City, CO.

Election Notice

The incorporation vote was set for May 25, 1892, approximately six weeks after the filing of the incorporation petition and the appointment of commissioners to supervise the election. The handwritten election notice was signed by each of the commissioners appointed to oversee the election—T. W. Vincent, John Gould, George C. Martindale, Frank H. Lara, and John Long. The legal document was prepared by the same Spencer & Watson, Attorneys for the commissioners. The election notice was accompanied by yet another handwritten note establishing that Ed O'Kelley had posted election notices in public places within Bachelor.

Finally, the package with the petition for incorporation contained three handwritten, sworn statements by commissioners John Gould, T. W. Vincent, and Frank Lara,[4] dated May 25, 1892. Each solemnly swore that: "I will perform the duties of Judge and Clerk according to the law and to the best of my ability: that I will studiously endeavor to prevent fraud, deceit and abuse in conducting the same, and that I will not try to ascertain nor will I disclose how any elector voted: if in the discharge of my duties as judge knowledge should come to me as to how any elector shall have voted unless called upon to disclose the same before some court of Justice." Commissioners Martindale and Long were not present during the vote, so sworn statements were not taken from them.

Vote for Incorporation

The final vote for incorporation was held, as announced, on May 25, 1892. It was the biggest day in the life of the young town. A total of 122 people, all males, voted. It must have been a lively vote because at the end of the day a significant number of people voted against incorporation. The ayes carried the day by a vote of 98 for incorporation and 24 against. This result was duly cataloged in the report by the commissioners who oversaw the vote.

A list of the 122 voters was included with the official declaration (fig. 22). Unlike the original petition, the names were handwritten in the same hand—probably by one of the appointed officials. How each person voted was kept secret, as required by the sworn statements of the commissioners.

List of Voters

1 Jacob Daum	2 A. M. Matthews	3 E. J. Marsell
4 Artha C. Joans	5 W. H. Correl	6 E. E. Steel
7 Jo McCabe	8 Frank Terrin	9 W. C. Sloan
10 E. J. Woodward	11 M. S. Steel	12 W. J. Roundtree
13 T. J. Ownsby	14 John Q. Day	15 H. V. McKinney
16 Z. T. Green	17 C. U. Coock	18 F. H. Lara
19 I. W. Newland	10 G. W. Stone	21 Ed Kelly (O'Kelley)
22 M. E. Calverd	23 J. F. O'Hanlan	24 Joseph Either
25 J. W. Granley	26 J. T. Mullinix	27 Frank M. Fisk
28 Fred Smitherin	29 R. T. Nelson	30 S. D. Fraser
31 A. M. Woodworth	32 S. T. Fenton	33 H. C. Gilmore
34 John Caldwell	35 B. Corkish	36 Jo. Kennaugh
37 Louis Hunkle	38 Riley Miller	39 J. C. Baler
40 C. H. Wood	41 Robt. Johnston	42 T. J. Green
43 Joe Lert [?]	44 A. R. Tetrault	45 S. Harrison
46 E. S. [?] Wells	47 A. Trimble?	48 V. McLind
49 C. A. Jones	50 E. L. McKay	51 F. L. Souva [?]
52 W. M. Hogue	53 J. C. Irwin	54 H. Larker [?]
55 J. W. Steel	56 James Francis	57 Randolph Good
58 M. H. Holt	59 W. J. House	60 J. M. Munson
61 Fred L. Kunry	62 C. Robinson	63 C. Simons
64 J. F. Grundy	65 W. W. Jordin	66 C. H. Duncan
67 O. W. Pennington	68 J. E. Snyder	69 R. Warthen
70 D. Mulcahy	71 W. W. McCoy	72 H. C. Conn
73 J. W. Jenkins	74 J. Pennington	75 H. C. Smith
76 S. Tognana	77 G. W. Buttler	78. B. Gist [?]
79 W. M. Waggoner	80 Wm. Kelley	81 M. McNulty
82 Wm. Ragen [?]	83 Clay Samuels	84 Henry Borne
85 R. F. Foreman	86 P. R. Claus	87 J. Born
88 A. C. Perry	89 S. D. Coffin	90 U. Moore
91 Chas. Wheeler	92 J. W. Deer	93 S. M. Funk
94 John Osborne	95 Frank Reed	96 W. C. Hoit
97 T. C. Ryan	98 J. W. Waren	99 M. Comerford
100 L. H. Brholoneu [?]	101 W. Streight	102 J. Kitchen
103 J. Sutton	104 J. G. Hogan	105 J. Hutchinson
106 J. P. Hawe	107 W. A. Ramriey	108 Lee Allen
109 J. Gould	110 T. W. Vincent	111 Patrick Moran
112 Terry Hughes	113 R. Heffron	114 D. Orr
115 W. M. Merritt	116 John Hannifin	117 Wm. Carter
118 Arom Van Canon	119 Wm. Burn	120 A. S. Crawford
121 C. S. Hall	122 Frank Love	

Tally List. For Incorporation – 98, Against Incorporation – 24.

Fig. 22. List of voters participating in the incorporation vote.
A question mark follows names whose spelling was uncertain.
Courtesy, Hinsdale County Museum, Lake City, CO.

As required by law, the results were published in the June 3, 1892, issue of the *Creede Candle*. The official announcement stated in part:

> *Said election returns so filed show that one hundred and twenty-two votes were cast therefor: ninety-eight of which were 'For Incorporation' and twenty-four of said votes were 'Against Incorporation.' Now therefore, I, Thos. J. McKenna, judge and acting clerk of said county court, pursuant to the statute in such case made and provided to publish and declare said above described premises duly incorporated under the name of the town Bachelor, and that, the same is an 'Incorporated town.'*[5]

To meet further legal obligations, L. H. Johnson, editor of the *Creede Candle*, filed a formal document with Hinsdale County stating that the notice of the incorporation vote had indeed been published in the June 3, 1892, issue of the *Creede Candle*. The document was dated June 4, 1892, signed by Johnson, and notarized by Alexander H. Major, notary public, Creede.

In the end, town residents had voted overwhelmingly in favor of incorporation. Colorado State and Hinsdale County law had guided the process, and a cadre of new leaders had stepped forward to ensure that incorporation became a reality. This historic event established Bachelor as a bona fide town. It also set in place a chain of events, such as the election of town officers that would further solidify the existence of a permanent town.

The list of people voting on May 25, 1892, is remarkable because it is the only written record of some of the males who lived in Bachelor at the end of May 1892. Women, who did not yet have the right to vote, and children were not included. The official U. S. Census would not take place until 1900, by which time the population of the camp had peaked and significantly declined. Many of the names on the 1892 list do not appear on the 1900 census.

The results of the incorporation vote were reported in a one-sentence article in the June 17, 1892, issue of the *Creede Candle*: "Teller, or Bachelor City, has been incorporated and an election of town officials will occur soon."[6] There is no evidence that the incorporation of Bachelor was publicized in the county seat of Lake City, despite the fact that a legal notice announcing the incorporation of Creede was published in the June 16, 1892, issue of *Lake City Times*. It noted that the incorporation election was held in Creede on June 3, 1892, with a vote of 507 "for incorporation" and 1 "against."[7] The author found no articles in the *Lake City Times*—from May–December, 1892—that mentioned the incorporation of Bachelor, which had taken place a week earlier than Creede.

On June 24, 1892, the *Creede Candle* had a short article with a list of nominees for mayor and the Board of Trustees of Bachelor:

The election for officers of the town of Bachelor will be held next Tuesday. The citizens have placed the following ticket in nomination: For mayor, Gustav Hoffman; for trustees, J. W. Jenkins, C. H. Pierson, I. W. Newland, A. S. Crawford, John Gould, A. H. Whitehead.[8]

Hoffman's name did not appear on the petition or any of the other incorporation documents. Frank H. Lara, George C. Martindale, and John Long, identified as "appointed" commissioners in the incorporation documents, were not on this list of nominees.

The report of the election of the slate of officers was listed in the July 1, 1892, issue of *Creede Candle*. The entire slate was elected, and the town moved forward to the next stage of self-governance: "The election at Bachelor Tuesday passed off quietly. The following officers were elected: Mayor, Gustav Hoffman; Trustees, J. W. Jenkins, C. H. Pierson, I. W. Newland, A. S. Crawford, John Gould, A. H. Whitehead."[9]

At their regularly scheduled meeting on Tuesday, July 19, 1892, Mayor Hoffman and Trustees Pierson, Gould, Jenkins, Newland, Likens, and Whitehead approved several actions.[10] The most important was passage of the sixteen ordinances that would form the legal foundation for the governance of Bachelor. The ordinances were approved and, by unanimous vote, the town officers passed a motion to have them published in the town's new newspaper, *Teller Topics*.

The vote of the trustees was reported in the first edition of *Teller Topics*, published on July 22, 1892: "Sixteen ordinances needed for the government of the city were passed. It was unanimously voted to have all ordinances published in THE TELLER TOPICS, consequently they appear in this issue and speak for themselves."[11]

Fifteen of the ordinances were printed in the same issue of *Teller Topics*. Ordinance fourteen was reprinted in the July 30 issue of *Teller Topics* to change the word "Creede" to "Bachelor" at the end of Section 1.

This suggests that the Bachelor ordinances were copied from or modeled after the Creede ordinances. A second ordinance fifteen was published in the August 6 issue of the *Teller Topics* (see Appendix 1). This was probably intended to be the sixteenth of the ordinances originally authorized by the Board of Trustees at the July 19 meeting.

The titles of the ordinances are:
1. An ordinance establishing rules and order of business.
2. An ordinance regulating the appointment of town treasurer and other subordinate officers, and prescribing duties not specially defined by statute.
3. An ordinance concerning licenses.
4. An ordinance relating to police magistrate, defining his powers, duties, and so on.
5. An ordinance concerning nuisances.
6. An ordinance defining the town seal.
7. An ordinance to prohibit the carrying of concealed weapons and provide punishment therefor.
8. An ordinance concerning road poll tax, and to provide for the collection of the same.
9. An ordinance concerning appropriations for the fiscal year beginning July 19, AD 1892, and ending on July 18, AD 1893.
10. An ordinance concerning fire—prevention thereof.
11. An ordinance concerning offenses against the public peace and to provide for the punishment thereof.
12. An ordinance to regulate the speed of driving or riding animals and to provide for the enforcement of the same.
13. An ordinance establishing the office of town physician.
14. Did not have a title, but constituted a board of health, composed of the town physician, town marshal, mayor, and board of trustees.
15. An ordinance concerning accounts. This is the first of two ordinances numbered fifteen.
15. An ordinance concerning dogs. Second ordinance numbered fifteen.

Fig. 23. Photograph of the tool used to make the Bachelor town seal shown in Figure 74. Courtesy, Grant Houston and Hinsdale County Museum, Lake City, CO. Photograph by author, 2016.

Each of the ordinances was signed by the new mayor, Gustav Hoffman, and attested by the town recorder, T. W. Vincent. The town seal, whose design is specified in town ordinance six, was affixed to each ordinance (fig. 74). The ordinances are listed in their entirety in Appendix 1.

The presentation and approval of town ordinances at the July 19 meeting was another important step forward in bringing order to the new town. As noted in the newspaper report, the Board of Trustees unanimously voted to have all the ordinances published in the *Teller Topics*. The ordinances provided a legal framework for how the town's business was to be conducted, listed the responsibilities of each member of the local government, addressed public safety issues such as law enforcement and fire protection, and provided enforcement mechanisms. The fact that the trustees felt it necessary for Bachelor to have ordinances supported the prevailing view that the new town would someday outpace Creede as the major city in the area.

The first meeting of the Bachelor Board of Trustees under the new ordinances was held on July 26 and the proceedings were reported in the *Teller Topics*.

The first meeting under the new ordinances was held on Tuesday night, July 26, at the town hall, with Mayor Hoffman in the chair. Present were Trustees Gould, Likins, Whitehead, Pierson and Jenkins, with Trustee Newland being absent. Following reading of the minutes of the July 19 meeting and other business, the mayor appointed the following standing committees for the fiscal year:

Finance—Whitehead, Jenkins and Gould.
Police—Gould, Pierson and Likins.
Streets, Bridges and Alleys—Jenkins, Whitehead and Pierson.
Fire Department—Pierson, Gould and Likins.
Journal and Printing—Gould, Pierson and Jenkins.
Health—Likins, Gould and Pierson.
Ways and Means—Whitehead, Jenkins and Gould.
The mayor appointed A. H. Whitehead mayor pro tem.

The recorder was instructed to order what printing was necessary for the town and its officers.

The marshal was instructed to enforce the poll tax ordinance and proceed to the collection thereof.

On motion a committee on ordinances was appointed. The chair appointed trustees Jenkins, Gould, and Whitehead as the standing committee.

Adjourned to meet in regular session Tuesday evening, August 3.[12]

It did not take long for the town to implement the new ordinances. A notice regarding collection of the poll tax was published in the same issue of the *Teller Topics* immediately following the report of the Board of Trustees meeting:

The marshal is instructed by the board of trustees to proceed to collect poll tax. It is hoped that no one will object at this time who can possibly pay, as the newly organized town mnst [sic] have some money to carry on its business. To avoid trouble we refer you to Ordinance No. 1, Secs. 1, 2, 3, 4. L. MacMullen, Town Marshal.[13]

Chapter 5
Boomtown

The summer of 1892 to the end of 1893 was an exciting time in Bachelor. This was the period of peak population and can be accurately described as the "boom period." The mines on the Amethyst Vein were undergoing explosive growth and new people steadily flowed into town. Mining equipment arrived at Creede every day, and there was a steady stream of wagons from Creede to Bachelor with a diverse supply of goods, from fresh produce to the latest in women's fashion.

The people of Bachelor were filled with enthusiasm and boundless energy. New businesses and homes were being built, and there was a sense of destiny. Many of Bachelor's leading citizens seriously felt the town was on track to become the largest and most important town in the Creede Mining District. In retrospect, these dreams were unrealistic because Bachelor was isolated and, most important, did not have a railroad. While rumors persisted that the D. & R. G. Railroad would soon extend to Bachelor, a sober assessment quickly dispelled them. Regardless, people believed strongly in the future, and it was a great time to be participating in the Bachelor boom.

Layout and Configuration of Bachelor

As mentioned, the town was laid out in a grid that had three named streets running north and south and nine cross streets. There were twenty-four blocks with twenty-four lots on each (see fig. 24).[1]

A panoramic photograph depicted earlier (fig. 1) shows that the town was populated along its entire length and breadth with businesses and residential houses.[2] Business buildings were scattered throughout the town, but the central part of the business district was on Main Street between Fourth and Sixth Avenues.

Fig. 24. Plat of the Bachelor City town site showing twenty-four blocks, with twenty-four lots in each block. North Avenue was the horizontal street at the top. The remaining streets were First through Eighth Avenues counting down (south). The vertical streets were Aspen (right), Main (second from right), and Park (third from right); the street farthest west (fourth from right) was unnamed. The plat, prepared by Field and McNutt Engineers, was part of the request to Hinsdale County for incorporation in April 1892. Courtesy, Hinsdale County Museum, Lake City, CO. Photograph by author, 2015

Forestry Service Road 504 runs up the east side of the town at the present time, roughly parallel to and/or coinciding with Aspen Street. The visitor's parking lot and informational sign are near the top (north) end of the town. The distance from the Creede visitor's center to the bottom of the town on Forest Service roads is 3.7 miles. A shorter and steeper road of approximately 2.5 miles was in existence in the late 1800s and early 1900s. The distance by foot, straight up Windy Gulch, is probably closer to two miles. The approximate distance from the Windy Gulch curve at the bottom of town to the top end, above the parking area, is 0.8 mile.

The residential sections were on Park Street, Aspen Street, and North Avenue, and extended well into the trees on the east, west, and north sides of the town. The photo in Figure 25, taken in December 1892, shows the town looking southeast across the main business section. Most of the panoramic views of the town were taken from the west because elevated rock outcrops served as perfect viewing platforms looking toward the east. The largest buildings were located on Main Street between Fourth and Sixth Avenues. This area had several businesses, saloons, boardinghouses, and liveries. It is the most photographed part of Bachelor.

Fig. 25. This December, 1892 view looks southeast across the middle part of Bachelor. This was the main business section, with residential housing extending to the east and west. Brooks and Drake, photographers, Amethyst, P.O., Fred J. Bishop Collection, Creede Historical Society Archives.

Fig. 26. View, looking east, of a portion of Bachelor around 1895. The houses,
probably on Aspen Street, extended well into the trees toward the crest of the hill.
The white business buildings, with large false fronts, can be seen in the center.
Made from glass negative, Harbert Archives.[3]

The town's water supply was provided by a spring. Harold Wheeler
describes it in his family history: "In the center of town was the pump,
where everyone got their water. What a job it was packing water and
keeping the horse trough full."[4] A pump from Bachelor is on display
on the outer grounds of the Hinsdale County Museum in Lake City,
Colorado (fig. 27). It is constructed from wood, roughly four inches by
four inches by seven feet high, with a tapered bottom and a hole in the
center for the water to flow. The handle near the top was used to pump
the water from the spout in the center.

Fig. 27. Wooden water pump from Bachelor that is on display on the grounds of the Hinsdale County Museum, Lake City, CO. Photograph by author, 2015.

The wooden sidewalks on Main Street were one of the main features of Bachelor's central business section. Harold Wheeler describes crawling under the sidewalks with his friends when he was a boy: "The wooden sidewalks were a real treasure trove. We crawled under them and found nickels, dimes, and quarters galore."[5] As can be seen in figure 28, the sidewalks were in front of nearly every business. In later years, some sidewalks were extended across the street (see fig. 52). They were one of the last visible features when Bachelor went into decline.

Fig. 28. An early (ca. 1892) closeup view of Bachelor's central business section on Main Street, probably between Fourth and Fifth Avenues, looking south down the street. The wooden sidewalks on both sides of the street were a key feature of this part of the town, which is not surprising given the muddy streets. History Colorado collection, Denver.

Bachelor Colorado 1908

Elevation: 10,500 ft.

To Creede

CLIFF

To Commodore Mine

N

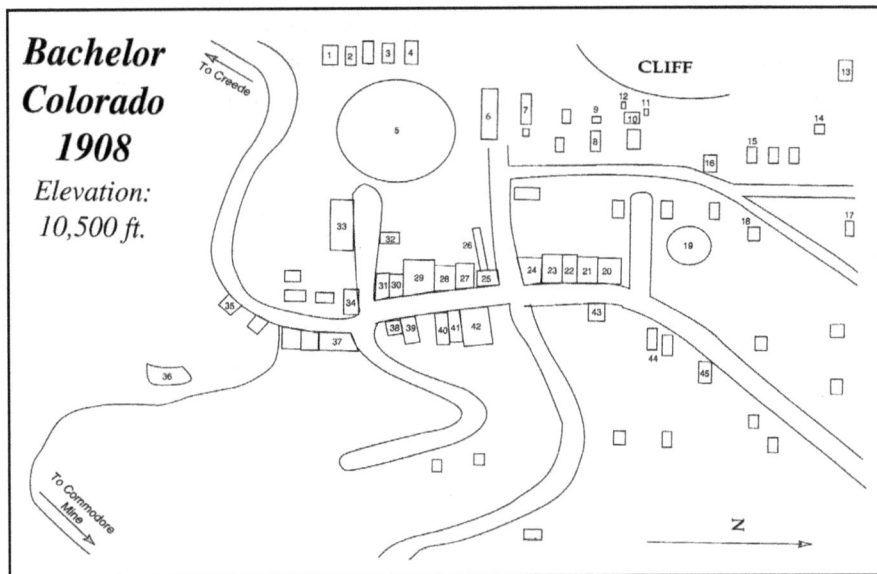

Fig. 29. Hand-drawn map—not to scale—of Bachelor in 1908 by Harold Wheeler. The map does not show all the streets but does cover the main business section and the residential area where the Wheeler family lived.

Key to Map of Bachelor

1.	Catholic Church	22.	Scotch
2.	Jail	23.	Cap Eades General Store
3.	Jenneve Lee, Harold's first girlfriend	24.	Barn
4.	Protestant Church with belfry	25.	Saloon
5.	Baseball field	26.	Bowling alley
6.	School	27.	Saloon
7.	Barn	28.	Grocery store
8.	McCoygue	29.	Dry goods store
9.	Backhouse	30.	Barber shop
10.	Wheeler house	31.	Saloon
11.	Woodshed	32.	Barber shop
12.	Backhouse	33.	Town hall
13.	Allen's house	34.	Liquor store
14.	Andy Uran	35.	Doc
15.	Jordan	36.	Cemetery
16.	Adams	37.	Fleming
17.	Pump	38.	Saloon
18.	Berry	39.	Boarding house
19.	Where Harold first saw Muriel, "the little red-headed girl" picking flowers on the hill	40.	Drug store
		41.	Meat market
		42.	Boarding house
20.	Burnt boarding house with stove	43.	Boarding house
		44.	Storage
21.	Boarding house	45.	Dunleavy

Fig. 30. Key to the map of Bachelor shown in Figure 29.

A hand-drawn map included in the Wheeler family history (fig. 29), while not drawn to scale, shows the relative positioning of some of the buildings in the central part of Bachelor. The town water pump, north of the central business section, is identified as icon 17 on the map.[6] The legend for the map is in Figure 30.

In its July 22, 1892, issue, *Teller Topics* states that: "Bachelor park is coming to the front as the bon ton residence portion of the city. It lays about a quarter of a mile up the hill, is finely timbered, and the view of the Rio Grande Valley is one of the finest to be had in the mountains."[7] This probably refers to the northern area of town, where houses were built at the edge of the surrounding forest (fig. 31).

Fig. 31. Residential area in Bachelor, ca. 1892, probably on Park Street or the next street west at the western edge of the town. Given the background and the slope down to the left, this was probably in the northwest part of the town, and may have been the upscale "Bachelor Park" area mentioned in the July 22, 1892, issue of *Teller Topics*. Photograph by Charles Goodman, who had a studio in Bachelor. Harbert Archives.

Elevation

The Bachelor town site has a commanding view of Snowshoe Mountain and the Upper Rio Grande Valley. The upper end of the town has a more expansive view, whereas the lowest part of town has a more spectacular view of the valley closer to Creede because the slope descends rapidly into Windy Gulch at this location. Bachelor's elevation has been listed in various references anywhere from 10,000 feet above sea level to above timberline (11,000–12,000 feet). The most accurate estimate was 10,500 feet above sea level.[8]

While walking the town site in the summer of 2015, the author found a U. S. Geological Survey marker (fig. 32) that lists the elevation at 10,531 feet above sea level. It was near the center of the town, close to where the central business section was once located. Given that the town is on a moderate slope, the lower end is approximately 10,430 feet, rising to approximately 10,630 feet above sea level at the top of the town.

Fig. 32. U. S. Geology Survey marker on the Bachelor town site. It lists the elevation as 10,531 feet above sea level. Photograph by author, 2015.

Population

The peak population of Bachelor was almost certainly in mid-1893, just before the crash of silver prices that marked the end of the boom period for the entire Creede area. It is therefore difficult to come up with an accurate estimate of the peak population because the town's first U.S. Census was not taken until 1900, nearly seven years after the end of the boom. An unofficial census was taken on April 13, 1892, by Ed O'Kelley and Joseph Either, during the process of submitting documents to support incorporation of the town (see fig. 20). O'Kelley came up with a count of 362, while Either had a count of 347. It is not clear whether these counts were to be averaged or should be added to each other. Regardless, the counts were taken well before the town reached its peak population over a year later.

Testimony of Bachelor's rapid growth in its first year came in the January 6, 1893, issue of *Creede Candle*: "Bachelor City, Teller postoffice, has grown in a year on top of Bachelor into a bustling business center of 1,200 population. It is one of the liveliest divisions of the camp and furnishes a trading place for 1,200 or more miners employed on the hill."[9]

The 1900 U. S. Census had a total population of 334, also well below the peak.[10] The highest estimated population, 6,000, was listed in the *1893 Colorado Business Directory*. This figure was certainly inflated, probably on purpose, to present a favorable picture of the town's prospects. Don La Font, who lived in the Creede area in the 1890s, described Bachelor as "a wild and lively and prosperous town of twelve to fifteen hundred people."[11] Without the benefit of accurate census information, this is a reasonable estimate by a person who lived there at the time.

We Need a Bank

In its first issue, *Teller Topics* lobbied for a bank in Bachelor: "One of the greatest needs of Bacholor [*sic*] at present is a bank. The merchants require such an institution and would gladly support one. Every pay day thousands of dollars could be handled to profit on exchange, cashing the checks for miners."[12] Some progress was reported in the next issue: "Two or three parties are trying to make arrangements to cash checks on the 1st. We trust some company will arrange for a bank, which from the start we think would be a paying enterprise."[13] The first step was the formation of a money order office, achieved in October 1892: "Bachelor is to have a money order office from to-day *[sic]* out. This is a convenience to our people that will be appreciated."[14] However, just a week prior to getting the money order office, *Teller Topics* was still highlighting the need for a bank: "Bachelor is the natural trade center for all the great mines of Creede camp. The payroll of the miners who trade at Bachelor is in the neighborhood of $1,000 a day. Give us a bank."[15]

The *Creede Candle* reported in its February 10, 1893 issue that Ellsberry and Foutch had added a banking department to its business.[16] The experience was short-lived, however, because that same year the company was caught trying to leave town without settling its accounts:

> *The Ellsberry & Fouch [sic] Banking and Mercantile company of Bachelor packed up their stock of merchandise on Sunday and attempted to ship all to Barton, Kansas, on Monday. Tuesday morning attachments began to come in from the outside. Sheriff Jones took charge of the goods and had them removed to the Wade & Hall building. The Miners Bank [in Creede] filed an attachment for $269; Daniels & Fisher $1,700; Cohn Brothers $850, and many others came in later. A transfer of stock had been made to the father of Foutch. The firm had done a large business in Bachelor but the hard times did them up. Foutch was treasurer of Bachelor. His bondsmen compelled him to turn over the sums in his hands.*[17]

It is uncertain whether another bank was founded in Bachelor. There are no bank listings in the *Colorado State Business Directories* for 1894–96.

Bachelor Has a Newspaper

The big event of summer 1892 was the publication of Bachelor's first newspaper, *Teller Topics*. Up until this time, news was provided by the *Creede Candle* several miles away in Creede. *Teller Topics* was the creation of John Shorten, a well-known Colorado newspaper man. The first issue appeared on July 22 with much fanfare:

> *THE TELLER TOPICS makes its appearance to-day, as it expects to for many Saturdays to come. No promises of future greatness or of immense achievements will be made. The paper is as you see it and will continue to interest and represent the business men and*

miners of Bachelor as long as the present friendly feelings between paper and town are maintained. The fact that the enterprise has been founded is proof sufficient that the publishers have faith in the permanency of Bachelor. We are here to help build up Bachelor in particular and assist in forwarding the greatest silver camp on earth in general. We hope to merit and receive your support in every way. JOHN SHORTEN, Editor.[18]

The paper advertised that: "Subscription rates to THE TOPICS have been placed within the reach of all. We want to spread the good word. Come in and give us a subscription for yourself and one for the folks back east. One dollar pays for six months and must be paid in advance."[19]

The layout of the paper was similar to that of the *Creede Candle*. The front page had a Mining Topics section that covered mining news for the Creede Camp. The fourth and fifth columns on the first page were devoted to Local Topics. Pages two and three usually had stories and news from outside the camp. Page four listed additional news pertaining to Bachelor. Advertisements were most prominent on pages one and four. Extra pages in the first edition contained town ordinances passed by the Bachelor Board of Trustees (see Chapter 4 and Appendix 1). *The American Newspaper Annual*, published in 1893–94, categorized *Teller Topics* as a Populist newspaper.[20]

The July 30 issue contained letters of encouragement from several local newspapers, including the *Creede Chronicle*:

Teller Topics *is a new journalistic venture in this district. It is the official journal of Bachelor City. The initial number has been received, showing business appreciation and support in the advertising columns and editorial ability and enterprise on the part of John Shorten and Lute Johnson publishers. It will be a welcome exchange.* —Creede Chronicle.[21]

Monte Vista Graphic added its praise:

The Teller Topics, *John Shorten editor, a new paper published at Bachelor, Creede Camp, is on our table. From the appearance of the first number it seems to be in the hands of men who know how to run a paper in a new mining town.* —Monte Vista Graphic.[22]

Shorten, who was also editor of the *Cripple Creek Herald*, had an interesting background. Born in Great Britain, he was compelled to leave because of his support of the Fenians, an Irish revolutionary group.[23] In the United States, Shorten saw service in the Confederate Army. He was well-known to have sympathies toward miners and their unions, and was arrested in Cripple Creek in 1894 for his support of striking miners.[24]

Shorten was heavily involved in the silver politics prevalent in western mining camps in the 1890s. However, in Bachelor he was president of a union with more fraternal aims:

> *The miners of Bachelor have organized a union, with benevolent and charitable objects. John Shorten is president, James Leonard vice president and Mr. Woodruff secretary. Eighty members are enrolled. It is emphatically announced that the organization will not undertake to regulate or govern the work of its members or tinker with the wage question.*[25]

The last issue of *Teller Topics* was published in 1894. *The Bachelor Sentinel* was published by C. O. Sprenger in 1894–95 and the *Bachelor Tribune*, T. W. Vincent, editor, was published from 1896 to 1903.[26]

Spring Creek Pass Road

Considerable excitement was created in July 1892 when a new mining district was formed on Spring Creek Pass, located between Bachelor and Lake City.

> *This camp, at the head of West Willow, twenty miles from Creede and up near timber line, is being gradually brought into prominence by well directed work on the strong leads which crop there.... The district is called Spring Creek and a townsite has been surveyed. There is quite a population of prospectors, and two ladies, Mrs. A. Z. Watson and Mrs. Thomas Wilson, are in camp.*[27]

This was important to Bachelor because a road from Bachelor to Spring Creek Pass would shorten the distance to Lake City, the county seat, and create a new business opportunity. The rationale for the road was highlighted in the September 10, 1892, issue of the *Teller Topics*.

Bachelor business men are nothing if not enterprising. The latest plan to increase facilities for miners of the camp which they have undertaken contemplates the opening of a good wagon road to Wilson's camp on Spring creek [sic]. The idea is to put through an eight-mile road from Bachelor. A line has been run and it is found that the distance to this new and bustling camp can be considerably shortened. It will also give a more direct route to Lake City, cutting off twelve miles from the present road.

A subscription paper has been circulated and over $100 is already assured. It is thought that there will be no trouble in completing it as most of the road is easy to build.

The importance of putting the enterprise through can be readily recognized. It will make Spring creek [sic] directly tributary to Bachelor and the miners there prefer to give their trade to our merchants.[28]

Progress on the road proceeded quickly, with completion expected in late September: "The road to Spring Creek, to connect with the saw mill road, will be open for business the end of next week."[29] Support from Lake City was also sought because the road benefited it as well as Bachelor and Creede: "The commissioners of Hinsdale county will be here soon to look over the proposed new road from Bachelor to Spring Creek. The citizens of Bachelor ask that $500 be expended at once and $500 next year, and the commissioners think favorably."[30]

The urgency of the project was emphasized the following week in an article in the *Creede Candle*:

A carload of ore is ready to ship from the Bondholder, in Spring Creek, as soon as the wagon road from Bachelor is completed. It is thought that that [sic] will be finished in a few days. Under the direction of T. R. Henahan the ore chute has been found, and there seems no question but that the Bondholder is to make a big mine.[31]

This was followed by an editorial in the *Teller Topics* praising the effort of the Bachelor community: "When it comes to rustle and enterprise, the merchants of Bachelor are strictly in it. Look at that Spring Creek road."[32] The road was apparently completed on schedule because this report appeared a week later: "The Bondholder shipped another car yesterday."[33]

Bachelor Businesses

Businesses quickly located to Bachelor once the town was established. In January 1892, the *Creede Candle* reported that there were already two saloons and the "female seminary" and that other businesses were expected.[34] The *Colorado Sun* reported that among the first buildings was a livery and feed stable built by C. H. Pierson (fig. 33) and a saloon run by a Mr. Carter. The paper proclaimed: "Stores and saloons opened with a rapidity only found in the West."[35]

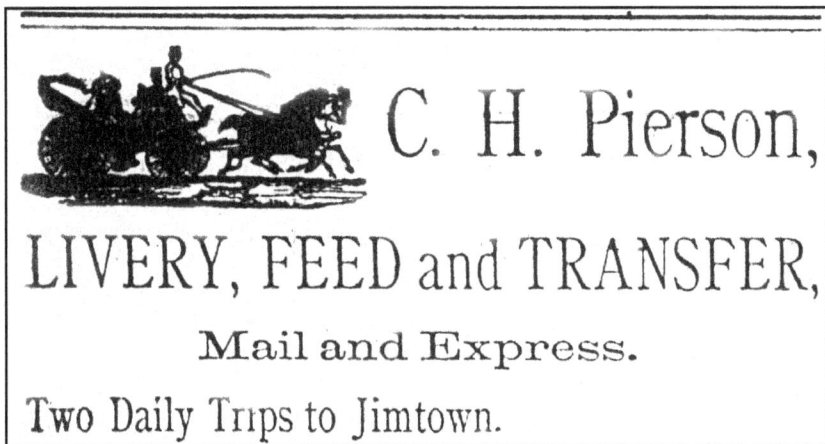

Fig. 33. Advertisement in the July 30, 1892, issue of the *Teller Topics* for C. H. Pierson, who established one of the first businesses in Bachelor.

Perhaps the biggest boost to local business was Bachelor's newspaper, *Teller Topics*. It provided a venue for advertisements and short articles in the "Local Topics" and "Around the Camp" sections. In addition, John Shorten, the paper's editor and publisher, was a big booster of local business. The newspaper published a steady stream of advertisements, articles, and brief notes to support local businesses. In its first edition, the newspaper stated:

Bachelor has the right kind of men at the helm. Her businessmen are of the right build and will make a 'go' of whatever they take in hand. We have no room to speak of each this week. Look at the adds [sic]. You will see a list of them and what they are doing. They are firm in their faith in the town and are backing up that faith with their works. Come up and be one of them.[36]

T. W. VINCENT,

City Recorder,

—AND—

Real Estate Agent.

General Business Transacted. Property Cared for.
Correspondence Solicited. Agent for
Bachelor City Property.

Fig. 34. Advertisement in the August 13, 1892, issue of *Teller Topics* for T. W. Vincent, real estate agent and the city recorder of Bachelor.

Pioneer Store.

Staple and Fancy Groceries

CLOTHING,

Hardware, Powder, Caps and Fuse.

First Class Meat Market in Connection.

JOHN GOULD, Propr.

Fig. 35. Advertisement in the August 13, 1892, issue of *Teller Topics* for John Gould, a founder of Bachelor and the first postmaster at the Teller post office.

THE TELLER HOUSE.

A. H. WHITEHEAD, Proprietor.

Good Day Board. First Class Accomodations.
Main Street, Bachelor.

Fig. 36. Advertisement in the July 30, 1892, issue of *Teller Topics* for A. H. Whitehead, one of the founders of Bachelor and a member of the original Board of Trustees.

Not surprisingly, many of the town's founders were also business owners who ran regular advertisements in *Teller Topics*. Several examples are listed in Figures 34–36.

The *1893 Colorado Business Directory* marked the first time Bachelor appeared in this annual publication that listed businesses throughout the state. The write-up at the beginning of Bachelor's entry, titled "Bachelor (Teller P.O.)" in bold print, is full of misstatements and hyperbole, including many misspelled words. Given the embellishments and bold statements about Bachelor's future, the piece may have been written by John Shorten, editor/publisher of the *Teller Topics*, who was a big booster of Bachelor's prospects:

Bachelor, Creede Camp, by virtue of its more favored location, is to-day [sic] the business center of Creede Camp. Beautifully situated in two of those famous mountain parks, each park being a step—one about fifty feet higher than the other—10,100 feet above the level of the sea, or about the same elevation of Leadville, which is 10,220 feet. The view from this town, over to the southern range and up the Rio Grande River for 100 miles is declared by travelers to be one of the finest on this continent.

Bachelor City is one of those happy combinations of chance and necessity so seldom to be found together when fate determines that a city shall be born. When that most remarkable body of ore that has ever been discovered, was opened to the world showing the necessity for the employment of thousands of men, and the transaction of millions of business, there, right at hand were those beautiful parks with their sparkling springs, ready and inviting, and Bachelor City sprang into existence as if by magic. The town site was surveyed in January, 1892, and in July of the same year the city was incorporated, with a full set of officials. The two parks, covering some eighty acres of ground, were covered with houses of all descriptions occupied by people of nearly every nationality. At this time, February, 1893, Bachelor is doing two-thirds of the business of Creede Camp. The post-office receipts indicating [sic] a population of about 6,000 people.

Bachelor has several first class hotels, a bank, several livery stables, blacksmith shops, hardware stores, grocery stores, confectioners, dry goods stores, bakeries, churches, Sunday-schools, dance-houses, a first class district school, saloons galore, lawyers, notaries, justices, woodhaulers, water venders [sic], newspapers and newsboys, bootblacks, and you can buy wieneworst [sic] and hot tamales [sic] at your door anytime. Bachelor has a miners' union organization in full blast and several of the secret organizations are represented. For so promiscuous a crowd, gotten together in so short a time, the town is comparatively quiet. It is almost useless to say that the voters of Bachelor cast a vote almost unanimous for the only silver candidate.

If anyone asks if this a mushroon [sic], or has Bachelor the elements of permanency, we only have to point to the results of the first year's development of the mines right at our doors, and when one remembers that the first year's work in a mining camp is work of development and preparation, the result is the most phenominal [sic] ever shown by any camp on earth—not excepting Leadville in its first year. The statistics in brief for Creede camp in its first year was [sic] that it showed up fifteen pay mines, there are twenty now. The number of cars of ore shipped the first year was 3,517 cars; tons, 46,355. Value five millions [sic] dollars. This output has increased every day until the result has startled every mining camp in America. We have the result for February, 1893, 9,000 tons amounting to $800,000, and all this from what is known as superficial development. Can the human mind conceive what the future of Bachelor City shall be when this vast body of mineral shall have been developed and it lays right at our doors.[37]

The unapologetic preamble represents the view, commonly held by *Teller Topics* and other leading citizens, that Bachelor was destined to surpass Creede and become the major town in Creede Camp. There were several valid arguments to dispute this position, but it represented the optimism of the time.

The directory itself includes sixty-eight listings, with saloons (nine) and hotels/boardinghouses (seven), grocery/butcher/bakery (seven), and clothing (seven) most prevalent. Some of the listings are duplicates of the business name and the individual(s) who owned the business; these were counted only once. Next in number of citations was construction/lumber (five). There were two physicians, two barbers, and three attorneys. Surprisingly, only one restaurant was listed, although hotels, boardinghouses, and saloons also served food. See Appendix 2 for the complete list of businesses in 1893 and succeeding years.

The number of businesses and people in the directory shows how rapidly Bachelor had built up. The directory was published in February 1893—only thirteen months after the town was platted and just seven months after it was incorporated. Indeed, it was prepared near the time when Bachelor had reached its peak population. By the time the 1894 directory was published, silver prices had crashed and the town was beginning its decline.

It was common for local businesses to pay for advertisements in their town's section of the business directory. However, Gustav (Gus) Hoffman had the only business ad in the 1893 directory (fig. 37).

Hoffman specialized in hardware and mining supplies, but he carried a broad array of merchandise, including cigars, stationery, and musical instruments. While he was not one of the residents who petitioned for incorporation or voted in the incorporation election, he clearly arrived on the scene early because he was elected the first mayor of Bachelor. From the photograph in Figure 38, it appears that his business was at the top end of the town, probably on Main Street.

GUS. HOFFMAN,

DEALER IN

HARDWARE, MINING SUPPLIES,

Clothing and Gents' Furnishing Goods,

Tobacco, Cigars and Notions, Stationery and Confec-

tionery, Musical Instruments.

Teller P. O.   BACHELOR, COLO.

Fig. 37. Gus Hoffman was the only Bachelor businessman to place an advertisement in the *1893 Colorado Business Directory*.

Fig. 38. Photograph of a business portion of Bachelor in the upper part of the town. The first business on the left was owned by Gus Hoffman, mayor of Bachelor. The Teller House, owned by A. H. Whitehead, was next door. The building on the right with the Zang Beer sign was probably a saloon. Fred Bishop Collection, Creede Historical Society Archives.

Fig. 39. Rare photograph of Hall's Restaurant, Bachelor. The view shows the business nestled among the trees, with a man and woman on horseback in front. John Gary Brown Collection, Creede Historical Society Archives.

Photographs of business establishments in Bachelor are fairly rare. The author has seen only five, including the photograph in Figure 38. The photograph of Hall's Restaurant in Figure 39 is the only photograph of a Bachelor restaurant he has seen.

As mentioned, hotels were among the most common listings in the 1893 directory. The Apex Hotel (fig. 40), originally the St. James Hotel, is an example. In its July 30, 1892, issue, *Teller Topics* announced: "Mr. Robert McKune, well known as a first-class business man, has leased the St. James hotel and will open on the first. The name of the house will be changed to the Apex, and under the new management will no doubt prove a success."[38] The following week the newspaper announced that:

Mrs. Ella Love of Jimtown will open a branch store in the building known as the St. James hotel on Monday, Aug. 1, with a complete line of stationery, all the current newspapers and magazines, also a fine line of confectionery, fruits, cigars and tobaccos.[39]

The newspaper still referred to it as the St. James Hotel, despite having announced the name change in its previous issue. Mrs. Love was listed in the *1893 Colorado Business Directory*, while the Apex Hotel and Mr. McKune were not. This is probably because the hotel had changed its name again to the Free Coinage Hotel (see Fig 42).

Fig. 40. The Apex Hotel and Mrs. Love's News Depot in Bachelor City, ca. 1892. A saloon is on the right in the same building. Additional businesses can be seen in the distance. The building was located on the west side of Main Street. Charles Goodman photograph. Harbert Archives.

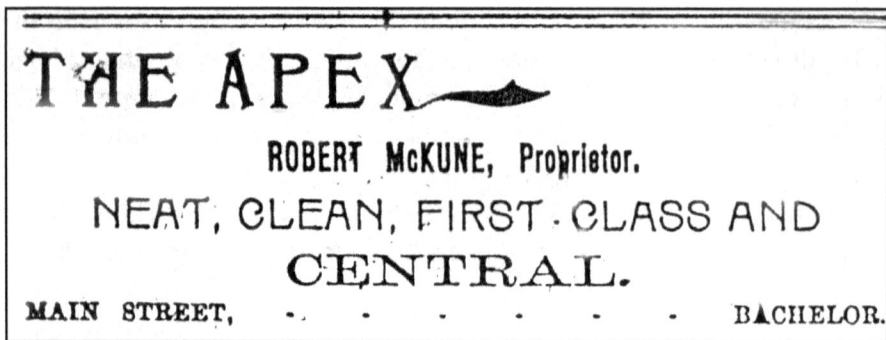

Fig. 41. Advertisement for the Apex Hotel shown in figure 40. It appeared in the August 13, 1892, issue of *Teller Topics*.

McKune took out an ad in the *Teller Topics* announcing the new name (fig. 41). The August 6, 1892, issue of the newspaper also announced that "Jay Finkelstein, the well known ladies outfitter and general dealer will be at the Apex House, late the St. James Hotel, with a full line of samples from the 6th to the 10th."[40] Continuing the trend of changing ownership, the September 10, 1892, issue of the paper announced:

George O. Martindale has purchased his partner's interest in the St. James Hotel bar. It will hereafter be known as the Denver Exchange. In this connection we must say that no better house can be found in the camp. It is well conducted, orderly and a place of resort.[41]

A photograph of the Free Coinage Hotel, taken in 1893, is shown in Figure 42. The two-story building appears to be the same building shown in Figure 40. If so, this was the third name for the hotel in two years. Mrs. Love's News Depot was no longer in the hotel when this photograph was taken. The Free Coinage Hotel appeared in the *1893 Colorado Business Directory*, and its proprietors were listed as Miller and Mitchell. The St. James and Apex Hotels were not listed in the 1893 directory. The Free Coinage Hotel burned down in 1894 (see Chapter 6).

As mentioned, the main business section was in the center of the town. Some businesses were apparently at the upper part of the town (see fig. 38), but an examination of photographs suggests that few were in the lower part of town. The white building in Figure 43 appears to be one of the rare businesses located there.

Fig. 42. Photograph of the Free Coinage Hotel, 1893. The lodging area was probably on the left and the saloon and restaurant on the right. Photograph by Hoer and Drake, Amethyst P.O., Sam Groves Collection, Creede Historical Society Archives.

Fig. 43. Photograph of a business (white front) near the bottom end of Bachelor, ca. 1895. The owner is probably the man wearing a vest and white shirt on the burro in front of the building. Given the many burros, it may have been a hardware or mining supply store. Creede Historical Society Archives.

Public Buildings

Once the town was established and growing, attention quickly turned to the erection of a jail, school, churches, and a city hall. At its first meeting on July 19, 1892, the Board of Trustees of Bachelor approved a contract for building a town jail. It was to be 12 x 16 feet, of 6-inch plank spiked firmly together, and containing two cells.[42]

One of the first major projects was the construction of a town hall: "The building of a public hall is being discussed by a number of Bachelor citizens and will probably materialize soon. It is intended that it be used for public meetings and also as a town hall by the Board of Trustees."[43] The building was constructed on the southeast corner of Park Street and Fifth Avenue.[44] In his memoir, Harold Wheeler asserts that it was better than anything in Creede and that all the big dances were held there. He mentions that the town always borrowed the Wheeler family piano for dances and that it was loaded onto a spring wagon and hauled to the hall. He notes that the fire bell on the hall was the same one later used on Creede High School.[45]

Construction of the Catholic Church was an early project and became a major fund-raising effort. It was first mentioned in the July 22, 1892, issue of the *Teller Topics*:

> *Father Downey said mass in town last Sunday and has undertaken to establish a church and regular services. He is meeting with much encouragement on all sides and is arranging to give a festival and ball on the evening of August 16 to raise funds with which to start the building.*[46]

The following week it was noted: "The town of Bachelor has donated a town lot to Father Downey on which to erect a church."[47]

The project moved quickly, and specifics were published just two weeks later: "Work on the new Catholic church is progressing rapidly and it will be ready for service by August 21st. H. C. Haynes is the contractor. The dimensions of the building are 20x40. It will hold 500 people and will be quite an ornament to our rising city."[48] Despite the article, one has to question whether 500 people could fit into a building this size.

A raffle to help finance the church was scheduled for August 16, and it was reported: "From the number of tickets sold in Jimtown and

Creede, we anticipate a large turn out on Tuesday night next at the ball to be given by the ladies of Bachelor for the benefit of the new Catholic church."[49] The fund-raiser must have been a success because maps of the town show that the church was located on Sixth Avenue, west of Park Street.[50]

Relative to all the publicity on the Catholic Church, little was said about the construction of the school (fig. 44) other than a brief mention in the July 22, 1892, issue of the *Teller Topics*: "Prospects are good for the building of a substantial schoolhouse in Bachelor before the opening of the fall term."[51] Examination of the Sanborn Fire maps from 1893 and 1904 suggests that it was on the west side of Park Street between Fourth and Fifth Avenues. The 1893 map shows a hall at that location, but the same building was listed as the school on the 1904 map.

There was also little information about the financing and construction of the Congregational Church shown in Figure 45. The handwritten map and legend in Figures 29 and 30 show the "protestant" church west of Park Street, several buildings north of the Catholic Church. It is possible that the Congregational Church was built later than the Catholic Church because a news article in the July 22, 1892, issue of *Teller Topics* reports:

Rev. J. B. Kettle organized a Sunday school at the schoolhouse last Sabbath afternoon. Forty-eight children were in attendance. Mr. Hayes was chosen superintendent and school will be held every Sunday at 2:30 p.m. Church will be held directly after the close of school each week.[52]

Whether Rev. Kettle was the Congregational minister is not known, but the use of the school suggests that the reverend was without a church at the time.

In his memoir, Harold Wheeler mentions:

The Catholic Church was down in the lower part of town. It did not have a bell, but the Protestant Church did. The bell tower stood just above the brow of a hill from our house. Just the right distance for a .22 rifle shot. Many a 'bong' I got out of it with my .22. This bell is now on the Church in Creede.[53]

Fig. 44. Bachelor Public School showing students posed for the photograph. Mrs. Laura Pollack, principal, and Miss Willie Caywood, primary teacher, are noted in the caption. The background is consistent with the suspected location on Park Street. Photographer unknown. Unknown donor, Creede Historical Society Archives.

Fig. 45. Photograph of the Bachelor Congregational Church, July 24, 1894. The background is consistent with the suspected location west of Park Street. The caption states that it is the highest-elevation church in the United States. Courtesy, History Colorado Photo Archives, Denver.

Finally, provisions needed to be made for a town cemetery:

> *One of the needs of Bachelor is a cemetery. Some of the citizens should get together at once and select a suitable spot. Mr. Simmons, in the south part of town, offers to donate ground sufficient for the property. No time should be lost to look over the location and make the arrangements necessary.*[54]

The cemetery was indeed located in the south portion of town. Given that the town sloped to the south and the spring was in the center of town, it was logical to locate it at the lower end.

Electric Lights in Bachelor

Creede got electricity in February 1892, but Bachelor had to wait because it was so isolated. In the meantime, it had to depend on kerosene lamps and candles for light. An early indication that electricity was on its way to Bachelor was published in September of that year:

> *There is a strong probability that in a short time Bachelor will be lighted by electricity. E. N. Magner of the Amethyst Electric Power and Light company was up here recently finding out from our merchants how many lights each one would take. The result, we understand, was entirely satisfactory. This company is preparing to put in an extensive plant to light all the divisions of the camp, the mines, and furnish electric power for all uses. It may seem advisable for the town board to stretch a few lights along Main street.*
>
> *Bachelor is coming steadily to the front as the business center of the great camp, and every new enterprise of this camp, and every new enterprise of this nature receives prompt encouragement.*
>
> *In all probability electric lights will be burning in Bachelor inside of sixty days. The town board has agreed to grant a long time franchise to the Amethyst Power and Light company, permitting it to erect poles and stretch wires in the town. At the next meeting of the board the ordinance covering this franchise will be passed.*
>
> *The company has gone to work to get up its plant. Superintendent Magner is now in Denver getting figures on the machinery and wires, and the ground will be broken for the power house and plant to be erected on West Willow in a few days.*

It is proposed to operate the machinery by water power. The plant will be an immense one, capable of supplying light and power for all the cities and towns of the camp, as well as drill power and incandescent lights for the mines.[55]

In December it was reported that the power company had changed management:

The Amethyst Electric Power and Light company has purchased the plant of the Creede company, controlled by the Poole brothers, and it is now managed by Dr. J. P. Wallace of the Nelson tunnel company.

This transfer removes the possibility of two companies competing for the supplying of lights in the camp.

Having secured franchises in Creede, upper Creede and Bachelor, the company has practically a monopoly in the business, but has developed no grasping tendencies up to date.

All night arc lights will be supplied at $18 a month, midnight lamps at $15. The company purposes [sic] putting in incandescent lights at once and will furnish these to all at $1.50 per month each.[56]

While this sounds encouraging, it is unclear when Bachelor received lights. An October 1894 article in the *Creede Candle* suggests that efforts were still under way then to get lights to Bachelor: "Dr. J. P. Wallace was in Denver this week buying a new boiler for the electric light plant. The old boiler will be sent to the Nelson tunnel plant. The electric company is enlarging to furnish lights to Bachelor and the tunnel."[57]

Examination of several photographs in this book under magnification shows no evidence of poles or overhead wires in Bachelor's early years. Poles are evident in photographs taken after 1900, so it is likely that electricity arrived in the mid– to late 1890s, but that is only an estimate.

Railroad from Creede to Gunnison

There was quite a stir in January 1893 when the possibility of a narrow-gauge railroad from Creede to Gunnison, going through Bachelor, was raised:

Creede promoters have undertaken a gigantic enterprise to furnish an outlet for the ores of Bachelor mountain and ultimately to connect the camp with the coal fields of Gunnison county.

It is nothing less than a complete narrow gauge railroad to run from Creede to Gunnison, via Bachelor City.

The preliminary section to be undertaken runs from the Amethyst mine to connect with the Denver and Rio Grande system at Creede station. This line is now being surveyed by C. A. King, city engineer. Three miles of the seven to be run have been staked at a uniform grade of 4 per cent. At this grade it will require seven miles of track to cover the two miles distance between the two points named. . . .

The projectors can not promise when work upon the grade will commence, but say the enterprise will be a go.[58]

Optimistic predictions that the railroad would be constructed appeared on a regular basis in the spring of 1893. A map of the southern portion of the Creede and Gunnison Shortline Railroad was published on the first page of the March 17, 1893, issue of *Creede Candle* (fig. 46). It showed the railroad line climbing Bulldog Mountain to Bachelor, then proceeding up West Willow Creek to the Continental Divide. A week later, another article stated that construction of the ore road to the mines was planned and efforts would soon begin to procure the right-of-way over mining claims. It also stated that a market for the bonds had been found and the capital was assured.[59]

In the end, the railroad was never built. Whether the failure of the railroad was a result of declining prospects for Creede mines, difficulties raising capital, failure to obtain suitable rights-of-way, or a combination of these factors is not known.

Relationship with Hinsdale County

Lake City's interactions with Bachelor were awkward at best. As mentioned, Lake City' leaders were unaware that Bachelor's post office was named Teller—just a week after the town was incorporated in Hinsdale County. In addition, the distance between the two towns— across the Continental Divide—was a major problem, particularly in the winter. It was natural for Bachelor to look to Creede rather than Lake City for commerce and other interactions. The *Lake City Times*

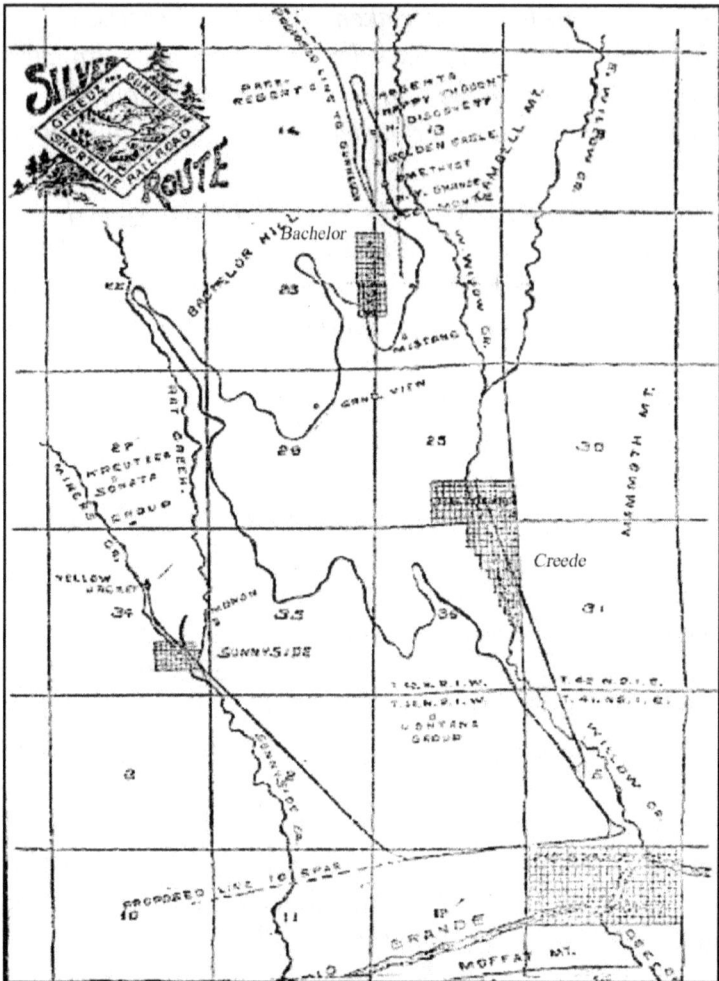

Fig. 46. Map showing the proposed route for the Creede and Gunnison Shortline Railroad, published in the March 17, 1893, issue of the *Creede Candle*.

recognized the problem and had to encourage county leaders to address issues in Bachelor:

> *The citizens of Bachelor are circulating a petition asking the county commissioners to expend a little money in extending and improving the road from Creede up to and beyond that flourishing camp. This section of the county has been ignored up to date by our county officials. Bachelor represents a large tax paying community in addition to being the greatest producing point in the county.*

Their interests should be considered without further delay and the road improved before winter sets in.

The above from the Creede Chronicle *would indicate that the residents of Bachelor are desirous of having better roads and The* Times *feels that the commissioners should at least look into the matter, and see that the roads about Bachelor and tributary points are equally as good as in any other portion of the county. Because Bachelor is stuck away off in the corner of the county is no reason why it should be entirely ignored.*[60]

Even with close attention and the best of intentions, the outcome was inevitable. A new county must be created.

Creation of Mineral County

As mentioned, the remote location of the Creede/Bachelor mining camp—at the intersection of Hinsdale, Rio Grande, and Saguache Counties (see fig. 14)—created serious problems of governance. This, combined with the rapid growth of commerce in Creede and Bachelor, constituted a compelling argument for the creation of a new county that could efficiently service local needs. Bachelor was clearly in Hinsdale County, but Lake City (the county seat) was fifty miles away over the Continental Divide, which could be impassable in the winter. Creede's status was more confusing because it overlapped county lines.

The idea of a new county occurred to savvy businessmen and politicians early in the mining camp's history. A December 1892 newspaper article suggests that planning was well under way at that time: "The committee to rustle cash to meet the expenses of the movement to secure a new county is meeting with good success. The merchants are coming down liberally and this proves that the effort has the indorsement [*sic*] of the solid men of Creede."[61]

Anxiety was evident in Lake City, Del Norte (Rio Grande County seat), and Saguache (Saguache County seat) over prospects of losing the rich mining camp. An article in the *Lake City Times* highlighted the quandary for Hinsdale County, which had the most to lose.

Last Saturday's Creede Chronicle *tells of an enthusiastic meeting held in Creede last Friday night, for the purpose of taking action in the matter of forming a new county, and making Creede the county seat. Officers were elected and committees appointed to push forward the work.*

This is a matter it is well for the people on this side of the range to ponder over seriously. Creede has a large number of hustlers among her population, men who will leave no stone unturned to accomplish their object, and the people of Lake City must make up their minds to sanction the move or else put on the war paint and prepare to oppose the scheme. If Creede takes a part of this county she will also fall heir to her share of the bonded indebtedness now hanging over Hinsdale.[62]

The *Creede Candle* began a steady drumbeat of editorials favoring the new county:

A straight forward business demand will be made on the incoming Legislature for the erection of a new county for Creede Camp. It will be backed by facts and figures to prove that a separate county government can be conducted, and there will be no excuse for any legislator opposing the measure.[63]

Hinsdale County residents eventually favored the creation of a new county. As mentioned, their major concern regarded the new county's acceptance of a portion of Hinsdale County's debt and the formula for calculating the new county's fair share:

The majority of Lake City tax-payers are perfectly willing that Creede should have a new county, the lines to be established in accordance with the bill recently introduced by Senator Timmons. Hinsdale people think that is the easiest way of any to pay from one-half to three-fourths of the $148,000 bonded indebtedness now hanging over them. By all means let Creede have her county, and the indebtedness, too, as the bill says the division of the indebtedness shall be made in accordance with the assessment of 1893.[64]

Mineral County was created in March 1893 from parts of Hinsdale, Rio Grande, and Saguache Counties (fig. 47):

Senator Timmon's bill for the creation of Mineral county passed the house of representatives on Wednesday morning by a vote of 56 to 6. It went through as amended in the committee of the whole, placing the county seat at Wason. Yesterday afternoon the senate concurred in the house amendments. The bill will be enrolled and may be signed by Governor Waite to-day or to-morrow. Mineral county will then be a fact and Wason a county seat.[65]

Bachelor and Creede were now in Mineral County. The county seat was moved from Wason to Creede on December 14, 1893.[66]

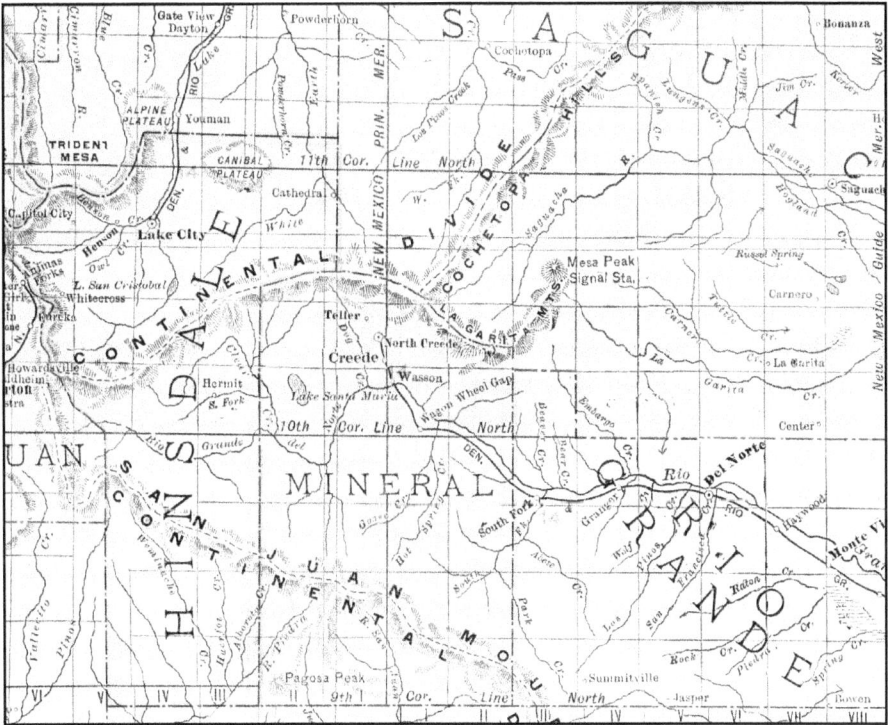

Fig. 47. Map showing Mineral County after it was created in 1893 from parts of Hinsdale, Rio Grande, and Saguache Counties. Teller (Bachelor), Creede, and North Creede are shown. Rand-McNally & Co.'s New Business Atlas Map of Colorado, 1904. Harbert Archives.

Chapter 6
Life in Bachelor

July 22, 1892 the *Teller Topics* reported: "The first birth in the town occurred Monday night. It is a girl, and was born to Mrs. Barngrover. *The Topics* is open to suggestions as to the most suitable testimonial to be presented to the young lady for her good choice of a birthplace."[1] The following week the newspaper stated: "By mistake in the first issue the first birth was credited Mrs. Barngrover. The first child born in Bachelor was a daughter to Mrs. Davis, and as the event happened some five weeks ago, the notice of last week is declared off."[2] Two weeks later the newspaper reported the sad story: "Died at 11 p. m., of indigestion, little Gorden I., infant son of C. L. and Amy Fuller, aged 2 months and 24 days. Funeral services were held at the house Wednesday, August 8, at 10 a. m.; burial immediately after."[3] In the same issue another story reported a mine accident:

> *Quite an accident occurred at the Equinox, close to the Cleopatra. Arthur B. Holliday fell a distance of twenty-one feet and broke his ankle. Dr. Biles attended him. He is now progressing. The accident occurred by the 'lazy pin' of the windlass breaking and was evidently the result of carelessness.*[4]

As in any western mining town in the 1890s, life in Bachelor ran the gamut of emotions from birth to joy to tragedy to death. The men working in the mines faced danger every day, and the confluence of young men, alcohol, and firearms meant the possibility of violence was always present. Fire danger was heightened by candles and open flames in structures built of wood and canvas. Yet at the same time art and culture flourished and children were able to experience the thrill of growing up in a pioneer town. And, there are even stories of humor in the lives of people living there. This chapter tries to capture the range of emotions, danger, and experiences present in Bachelor.

Violence

Violence, including gun-related deaths, was common in western mining towns, particularly during their so-called boom days. Creede gained a reputation as a "wide open" town, and Bachelor developed a pattern of drunkenness and violent outbursts that rivaled that of the larger town. The *Creede Candle* asserted: "Bachelor City is one of the liveliest of the many towns in the camp."[5] The statement didn't just have violence in mind, but it was an apt description. After a July 1892 weekend of drinking and violence in the town, the *Teller Topics* editorialized: "Bachelor cannot afford to have her reputation become that of a tough town. The officers are acting promptly and forcibly and must have the moral support of the community."[6]

Caroline Bancroft wrote that Bachelor had competent law officers and the support of its leaders:

> But the efforts of the better people failed. The character of Bachelor remained tough. At the height of its population of around twelve hundred, two hundred were prostitutes. It was a nightly custom for patronage of the soiled doves to include not only the local boys, but miners from Creede, North Creede, and Weaver, who tipped the hoistmen of the Last Chance and Commodore to lift them up to the wild, brawling and drunken delights of Bachelor.[7]

A striking example of the danger of guns, testosterone, and liquor occurred in June 1892 and was captured in the *Creede Candle*. It illustrates how quickly things could get out of hand:

> Thomas Coyne was fatally shot by Bill Hogue at Bachelor City last Sunday night. Coyne, with three or four others, had been drinking and carousing all the day previous and were bent on running the town in their own peculiar style, and were succeeding to their entire satisfaction, but greatly to the discomfiture of the inhabitants. They proceeded to make things generally lively at the dance hall in the evening, and when requested to desist by Sheriff Soufa disputed his authority to interfere and a wrangle ensued, in which Coyne was taken by the sheriff. He was handcuffed and turned loose to appear for trial the next morning.

Assisted by friends, he succeeded in cracking the chains and, accompanied by his associates, proceeded Sunday evening to 'look up' the sheriff and have a 'reckoning.' Soufa, learning of the gang's intentions, proceeded to arrest Coyne, whom he found in a saloon. Coyne resisted and was assisted in his escape by one McCoy. The sheriff took McCoy in custody and deputized Bill Hogue to arrest Coyne. Hogue, with a shotgun, followed him to his cabin and attempted to arrest him. Coyne showed fight and two loads from the gun were emptied, one taking effect in the hip and groin and one grazing his head.

Coyne's friends, learning what had happened, went in search of Hogue with a rope in their hands and blood in their eyes and it looked for a time as though other lives would be lost. However, when Hogue was found by them he proved to be armed to the teeth and prepared to sell his life so dearly that his pursuers were loth [sic] to tackle him. Three of the most determined made a break to secure him and as many were laid low from the clubbed shotgun in his possession. He then held the crowd at bay with a revolver in each hand until he made good his escape from the angry mob and finally through assistance from friends got out of the camp.

Coyne's wounds are very serious and the doctors say that he cannot recover.[8]

In a follow-up story a week later the *Creede Candle* reported:

Thomas Coyne, who was shot by Bill Hogue at Bachelor on the night of Sunday, June 19, while attempting to escape arrest, died at 2 o'clock Tuesday afternoon. Hogue is in jail at Del Norte held under $1,000 bonds. Coyne received two loads of shot in the back at the time of the shooting, but was not thought to be in serious danger until within a short time before his death.[9]

Another shooting incident in Bachelor in July was reported in the July 29, 1892, issue of the *Creede Candle*. Fortunately, this "affray" was resolved without loss of life:

Bachelor was the scene of a lively shooting affray Sunday night. Mike Donnelly was joshing William Gardenheir after the latter had interfered in a conversation the former was holding with a friend *at the St. James hotel bar. When Gardenheir retorted, Donnelly smacked his mouth. Gardenheir went after his gun and, returning, opened fire. He had shot three times before Mike got his gun into play, and then the cartridge missed fire and Mike started to get out of the way, going around the hotel, pursued by Gardenheir. Mike got his gun to working about the time the two reached the back door of the barroom and afterwards used it to club his assailant's head. Of the many shots fired in the melee, Donelly [sic] caught one ball in the shoulder and Gardenheir received one in the right arm and one in the side, besides a badly smashed head. The shooting attracted a great crowd. Marshal MacMullen was promptly on the ground and stopped the row. Gardenheir was taken before Justice Beatty and placed under $2,500 bonds. Neither party is seriously wounded.*[10]

While the fight was going on, "Kid" McCoy, who was in jail at the time, decided to take advantage of the gunfight. As reported in the *Creede Candle*: "'Kid' McCoy was arrested Sunday for disturbing the peace of Bachelor citizens and [was] confined to the town jail. During the shooting on Sunday afternoon he set fire to the jail, but the flames were discovered in time. He was placed under $500 bonds."[11] McCoy was the same person arrested in the Soufa/Hogue/Coyne fracas described previously.

Marshal MacMullen was praised for his handling of the Gardenheir affair in the July 30 issue of the *Teller Topics*: "The prompt action of Marshal MacMullen during the difficulty on Sunday night was commendable. He showed himself to be the man for the place."[12]

Conflicts between miners and mine owners were common occurrences, but one such altercation in Bachelor escalated to violence and resulted in the accidental shooting of a bystander. In this case, the mine owner was Albert E. Reynolds, a major investor in Colorado and Creede mines. Things got out of hand when the miners physically attacked Reynolds:

The town of Bachelor, near Creede, was the scene of a tragedy on the 22nd, by which John Erskine lost his life.

There had been trouble between the miners on the New York and Chance and the manager, A. E. Reynolds. He had closed down the mine because the men refused to allow non-union men to work at $3 for a ten-hour shift, and there was a good deal of excitement in the town. At 3 o'clock in the afternoon Kid McCullough, one of the workmen locked out, accosted Mr. A. E. Reynolds for his pay check. Mr. Reynolds offered to give the check, but wanted to hold out the miner's board bill, as is the rule. At this McCullough slapped the manager, and his brother Harry poked a gun into his face. So threatened Mr. Reynolds gave a check for the full amount.

The McCullough boys then went to their boarding house, the City hotel, in Bachelor and got into a dispute with Mrs. Marshall over the bill. Frank Everett, shift boss at the New York, and one of the proprietors of the house, interfered. Kid McCullough kicked over a table and a general fight was in progress when City Marshal Charles Duncan appeared and attempted to stop the row. At this the McCulloughs attacked the marshal. Kid pulled a gun and Harry beat Duncan over the head with a club. Duncan took the gun from Kid and fired. John Erskine, a young man and an innocent spectator, was fatally shot. Five shots were exchanged.

Duncan was distracted with grief and unable to give any account of the trouble. He went down to Creede and gave himself to Under Sheriff Gardner. It is thought that more trouble may be expected as the men are worked up to a high pitch by the incidents of the day.

Mr. Reynolds sent for the miners and told them that all who wanted to go to work at the regular time in [illegible] and that wages would be $3 for ten hours. He promised to pay the day men for the last shift if the men went to work that night, but if they did not he would shut down the mine until the owners could run their own property, even though it remained closed for ten years. Many of the men went to work under this understanding, and it is probable that the labor trouble is over.[13]

The article was published in the April 1, 1893 issue of the *Silver Standard* in Silver Plume, Colorado. A week earlier, two articles in the March 24 issue of the *Creede Candle* provided background and a slightly different perspective. The latter newspaper reported:

Intestine troubles between the workmen employed in the mine led to the shutting down of the New York and Chance on Wednesday morning. Before the day closed the misunderstanding which *precipitated the action had been removed and the mine re-opened, but the day was frought [sic] with events that will make it long remembered.*[14]

Regarding the gunfight, the same issue of the *Creede Candle* reported that Duncan was nervous and rattled because he had been sworn in only a few hours earlier. While the story in the *Silver Standard* described Erskine as an innocent bystander, the *Creede Candle* stated:

Another story is that Erskine was clubbing Duncan over the head with a billy: that the first shot from Duncan's gun hit him in the heart and caused him to fall forward with his hands on the marshall's [sic] shoulders in which position he received three more balls. Erskine fell to the floor, gasped a few times and died. A formidable billy was lying by his side when outsiders came upon the scene.

The article continues:

After shooting Erskine, Duncan turned upon Kid McCullough, beat him over the head with the captured gun, arrested him and locked him up and then came hurriedly to Creede and gave himself up to Undersheriff Gardiner. . . A coroners inquest was held in Bachelor yesterday afternoon. The verdict declared the killing to be unwarranted but assistant district attorney Strickler, after hearing the case against Duncan, holding [held] that he was fully justified and acting in his own defense.[15]

Another tragic incident began as a dispute over, of all things, a turkey raffle. The initial conflict was reported in a somewhat humorous manner in the March 3, 1893, issue of the *Creede Candle*:

Because two of Bachelor's citizens didn't agree as to the way a turkey raffle terminated, the rarified atmosphere thereabout was recklessly perforated with bullets Tuesday evening. And because of such perforations, the justices of Creede have been busy about all week dispensing legal salve to the peace and dignity of the people of the State of Colorado. H. W. Woodruff, colored, let off a few shots and Judge Jennings considered the fun he had worth $20 and costs, and so assessed. *The charge was for assault to kill, but the judge decided Woodruff had no malice in his heart and so made it plain assult [sic]. Michael Sherry was the other party to the complaint. He was tried before Justice Whitely and a jury yesterday and acquitted.*[16]

Unfortunately, tempers simmered after the affair and ended in violence with clear racial overtones, as reported one week later in the *Creede Candle*:

W. H. Woodruff was shot and instantly killed by Michael Sherry in front of the Palace bar at Bachelor at noon of Monday.

As told in these columns last week, a difficulty between Woodruff and Sherry arose over a turkey shoot. Woodruff at that time appeared to have been the aggressive party and was fined $20 and costs for assault. Sherry was cleared of a like charge.

After the trial was over Woodruff returned to Bachelor and from reports had repeatedly threatened to "do up" Sherry at first opportunity. Monday the two men met for the first time since the trouble. Sherry had heard the threats made by Woodruff and as soon as he saw his adversary pulled a 38-calibre gun and opened action. He fired three shots, the first taking no effect, the other two entering Woodruff's brain and killing him instantly.

Sherry at once hid out and spent the night in a cabin at Sunnyside. He came back, however, Tuesday morning and gave himself up. He is held without bonds, having waived examination.

Woodruff is a colored man who has been in Bachelor since the early days. He has conducted a laundry and bath house business in which he was successful. Although at times a bluffer and inclined to quarrel, he was generally attentive to business and well-behaved. He was buried with a soldier's honors in Sunnyside yesterday morning.

Sherry is one of the proprietors of the Miners' Home Saloon and has a host of friends all over camp. He is a much smaller man than the man he killed, but is full of nerve and doesn't like to be crossed. He had been jeered for standing the attack of the negro and this was probably as much the cause of him killing Woodruff as was his fear of a second attack from him.[17]

Not all stories of conflicts with Bachelor law officers ended in violence. In fact, some were humorous and did not result in death or injury:

Two cow punchers, fresh from the range, struck lower Creede Thursday and proceeded to store away six months back rations of pine-knot under their pistol belts. When the seductive fluid began to get in its work, some friends of theirs with a terror of the laws and the police court in their hearts steered the pair up to Bachelor, where it was supposed they could blow off their exhuberance [sic] without danger of arrest. They came into town having plenty of fun, making echoes on the light air, and thought it would be funny to joke the people on not having a marshal to pull them in. 'Where is your —— [expletive] marshal?' they yelled. Now, it came to pass that the upholder of the peace and dignity of Bachelor stood rubbing shoulders to the propounder of the question. Marshal McMullen is nothing if not accommodating. Seeing he was wanted, he stretched out his arm and, laying it tenderly on the disturber's shoulder, smiled sweetly and said, 'Here I am sir.' It cost the pair $85 in fine and costs to find out that Bachelor is incorporated and not only has a marshal, but a rattling good one.[18]

Another violent confrontation occurred in 1905 between two mining partners:

Late last evening Andy Wellington went hunting trouble at the home of his partner, in mining claims, A. R. Allen, in Bachelor and was speedily disposed of. According to Allen's statement he was in the back yard sawing wood when Wellington approached him and threatened violence. Allen got away and his wife and daughter undertood [sic] to protect him from further harm when Wellington struck and knocked both of the women down, breaking the front teeth in the mouth of the girl. Wellington then went after Allen again who had hurried into the house and armed himself with a double barreled shot gun and went out the front door and as he looked around the corner of the house Wellington fired and Allen carries a slight flesh wound on the upper lip midway between the nose and left corner of his mouth, which he says is where the bullet struck him. Allen then instantly drew up his gun and fired and Wellington fell dead. There were no witnesses to the affair other than Allen and his family, but the report of shots brought most of the citizens of the town to the scene where they found Wellington dead with the smoking bull dog pistol by his side. Allen was taken in custody by officers and Wellington's body sent to the morgue and upon examination disclosed a terrible looking sight. His right side and arm is literally covered with buck shot holes and around from the right nipple is a large hole evidently by a big slug which passed through the body and gives evidence that the gun was loaded for some special occasion.

Allen says that he and Wellington had never had any trouble and was surprised at this attack, but some of the residents state that there was some altercation between them either over a claim or one of Allen's daughters with whom Wellington has been keeping company for that past year and the supposition was not long ago that they were to be married.

Evidence undoubtedly proves that Wellington premeditated the attack and evident killing of Allen for on investigating his cabin after he was killed it was found to be in readiness for an attack. His rifle, loaded and even cocked was laying [sic] upon the bed and numerous cartridges laying [sic] about in convenient places. Wellington was mentally deranged of which he gave particular evidence at times of being provoked and had served one term in the penitentiary for attempting to kill Taylor Markley while in one of these tantromes [sic] and was liberated from the state prison about three years ago.

District Attorney Pilcher came up this morning to investigate the deplorable affair and this afternoon the inquest was held in Bachelor and the jury found that the killing was justifiable homicide.[19]

The most famous murder in the Creede Mining District occurred on June 8, 1892, when Ed O'Kelley murdered Bob Ford in the latter's saloon. Ford was famous for having killed Jesse James, the notorious bank robber from Missouri. While Ford's murder took place in Creede, it is mentioned here because O'Kelley was one of the founders of Bachelor and was the town marshal at the time. He was one of the original thirty-three petitioners for incorporation on April 13, 1892, and was among the 122 residents who participated in the incorporation vote on May 25, 1892. The June 17, 1892, issue of the *Creede Candle* reported that

"Ed Kelly [sic], the murderer of Bob Ford, was taken to Del Norte on Monday for confinement in the Rio Grande county jail. Joe Either, the accomplice, was taken on the same train."[20] Either, also from Bachelor, conducted the census with O'Kelley as part of the town's petition for incorporation (see Chapter 4). While there are various accounts of who helped Ed O'Kelley kill Ford, the author believes this is the first book to definitively identify Joe Either as the accomplice.

Another example of a Bachelor town marshal gone astray occurred in the spring of 1893:

Harry Conway, town marshal of Bachelor, was arrested Saturday afternoon, charged with having in his possession some of the jewelry taken from the pawnbroker's store of Levy & Sugar, in South Creede, September 10. The place was burglarized and $2,000 in watches and jewelry taken, together with $200 in money. Dan Collins is also under arrest charged with being implicated. Conway was given a hearing before Justice Webb and fined $20 and costs for having in his possession stolen property and not exceeding in value $15. Collins was bound over.[21]

There were constant fights in Bachelor, usually because men had too much to drink, but one of the funniest stories involved a 1906 fight between two women. The story was told by Edwin Lewis Bennett, co-author of *Boom Town Boy*, who grew up in nearby Weaver and delivered mail between Creede and Bachelor:

I saw two fights in Bachelor that spring and each was odd in its own way. The first was not between men but between two women, one of them Irish and the other Cornish. They had been quarreling at each other for some time and, coming downtown that day, had run into each other and started jawing. Their husbands, fed up with the long feud, agreed that was the time to get it settled so they made the wives fight it out, Marquis of Queensbury, without any scratching or hair-pulling, but man style. Foster's saloon was at the upper end of town, and the fight took place right out in front, so we had a ring-side seat. Occasionally one of the women would revert back to type and bare a claw or get a handful of hair but her husband would make he [sic] back up and start clean again, so it was a nice, respectable battle. There were no rounds. The women were both fairly well padded and short-winded, and the time came when they were panting and taking wild, aimless swings at each other. As one had the makings of a good black eye and the other had a bloody nose, their husbands thought they ought to have it all out of their systems and stopped the fight. The battlers sat down on the bench in front of the saloon to rest and get their breath, and, one of them happening to mention that she had some beer on ice up at her house that might do them both good, they went there, leaving their husbands to get the groceries they started after. After that fight each of them had one more friend than she had before and the husbands didn't have to listen to any more name-calling.[22]

Fires

Compared with Creede, Bachelor had relatively few serious fires. In fact, "After the fire in Creede on June 5, 1892, many of its citizens went to the high mountain townsite of Bachelor, more than doubling its population."[23] This reflects the view that Bachelor was relatively unscathed by fire. However, it is unlikely that enough people moved from Creede to Bachelor to double the latter's population.

Two fires occurred on January 11 and 12, 1893. While property was destroyed, neither fire resulted in loss of life:

> *The city on the hill had two narrow escapes from distruction [sic] by fire on Wednesday night.*
>
> *At 10 o'clock flames were discovered pouring from the log bunk house of the Bachelor mine. The people turned out and fought like Trojans with the result of saving the town with the loss of only a few cabins. the [sic] oldest house in the camp, was burned. Three other log structures were torn down to stop the fire.*
>
> *The reflection was plainly visible in Jimtown and now and then the anxious watchers in the gulch could see flames dart skyward. Much uneasiness was felt for the fate of Bachelor by the citizens here and the absence of definate [sic] information gave rise to rumors of its total distruction [sic].*
>
> *Another fire broke out at 5 o'clock Thursday morning in the cabin of Lone and Finley. Six men were sleeping in the place and being aroused by the smoke gave alarm and assisted in getting*

the fire under control. Nearly all the contents of the cabin were distroyed [sic].[24]

A little over a month later, on February 26, fire struck again: "The Last Chance saloon and Hilton's photograph gallery at Bachelor were totally destroyed by fire Sunday night. The former was insured."[25] E. F. Hilton was one of the area's best-known photographers. While his main studio was in Creede, he also established a studio in Bachelor as the town began to grow.

In an effort to improve Bachelor's firefighting capability, the town invested in a new fire engine. As reported in the May 5, 1893, *Creede Candle*:

> *A test was recently made in Bachelor of the new chemical engine, furnished by Miller & Marvin. The engine has a capacity of sixty gallons of chemical fluid, which is equal to 800 gallons of water. The trial was perfectly satisfactory, and as a consequence the citizens of the town in the clouds rest easier. They have a well-trained fire company to handle the machine and fight fire should one break out, and in this particular are better off than their neighbors in Creede.*[26]

The new chemical machine was used without success against Bachelor's most destructive fire. It occurred when the Free Coinage Hotel (fig. 42) burned in January 1894:

> *At 1 o'clock Saturday morning the Free Coinage hotel at Bachelor was discovered to be on fire, by flames issuing from the rear end. An heroic fight was made with the chemical engine and while effective work was done in preventing the spread of flames, the fire had taken such a hold on the big frame hotel that nothing could save [it], and it burned to the ground, together with the old Lundy & Sherry saloon building adjoining.*
>
> *T. W. Vincent, chairman of the board of county commissioners and J. Thatcher are the principal losers, they having owned the block destroyed. The building was one of the first to be put up in Bachelor and was the largest in town. It cost $5,000 to build and was insured for only $1,500. Pat Collins had a saloon in the building on which he carried $500 insurance.*

No one was in the building when the fire started and it is not known how it originated, but the loss is attributed to a defective flue.[27]

One of the disadvantages of the Bachelor location was that it was surrounded on three sides by spruce, pine and aspen trees. As such, the town was vulnerable to a forest fire encroaching on the town site. Such a scare occurred in the summer of 1893:

At about 10 o'clock this morning word came down from the hill that the town of Bachelor was on fire, having caught from the timber fires all around. In a few minutes half the population of Jimtown was headed up the hill to lend our neighbors all assistance possible. Every horse and vehicle was brought into requisition; Hoover sent everything in the barn and many went a'foot. Luckily the report was exaggerated. The fire burned to the edge of town but a most dogged fight on the part of citizens saved the town with the loss of only a cabin or two.[28]

A bizarre case of suspected arson occurred in February 1893. It was unusual because the suspected arsonist was A. A. Vance, one of the founders of Bachelor who signed the original request for incorporation:

There was a great sensation in Bachelor Monday morning. On Saturday morning an unsuccessful attempt was made to burn the Apex Hotel [fig. 40]. Letter paper having the name of A. A. Vance printed thereon and a written order to his brother-in-law, Mr. Johnson, was [sic] found saturated with kerosene. Suspicion in this way pointing to Johnson, he was approached by the citizens in a threatening way and gave evidence that seemed to implicate Mr. Vance. A delegation visited Vance with a rope and belligerent aspect, but he was allowed to be placed under arrest. He was to have had a hearing in the city on Tuesday, but the case was postponed for a week. Everybody seems to think that Vance is guilty.[29]

Vance escaped from Creede, and a month later this statement appeared in the *Creede Candle*: "Undersheriff Gardiner returned to town on Wednesday with A. A. Vance charged with attempted incendarism [sic] at Bachelor, who escaped from deputy Comstock several weeks ago.

Vance was at Oswego, Kansas, working on a farm."[30] Vance must have been an escape artist because an article in the *Creede Chronicle* a few months later provides the next chapter in the saga:

Under Sheriff Gardiner, at considerable expense to the county captured A. Vance, charged with incendiarism at Bachelor, last winter. Vance was taken to Lake City for safe keeping, and thence to the Gunnison jail. He escaped this week, and justice and Mineral county are short a desperate prisoner. District Attorney Merriman says the County should have a jail, where prisoners can be confined without taking the risk and increasing the additional and unnecessary expense of transporting them from one county to another. There is no reason why the circumstances should not give this matter early attention.[31]

On the front page of the same issue of the *Creede Chronicle*, an editorial argument was made for Creede having its own jail: "A. Vance, the Bachelor firebug, escaped from the Gunnison jail this week. The county commissioners should provide the sheriff with a jail in the county and do it without regard to what may or may not happen in the sweet bye-and-bye."[32]

Anyone who has been in the high country around Creede in the summer is fully aware of the danger of lightning. Bachelor was not immune, as this story from 1912 illustrates:

Word was brought down from Bachelor Monday that during the electrical storm of that morning, a bolt of lightning had descended and striking Bud. Parkinson's cabin, had set it afire. The cabin and all of its contents was [sic] utterly destroyed, all that was saved being a cross-cut saw. The storm, although of brief duration, was very heavy while it lasted.[33]

Growing up in Bachelor

Some of the best stories of Bachelor were told by people who grew up in the town. One was Harold French Wheeler, who moved to Bachelor with his family in the early 1900s. In a memoir written for his family, he describes his experiences growing up in the town. He and his friends had considerable freedom to roam the hills and mines around Bachelor, and his stories are full of fun and potential danger. His granddaughter, Susan Weston, wrote the introduction to the memoir:

> *Harold Wheeler was the inspiration for this search into family roots and oral history by his granddaughter Susan. In the months prior to his sudden death in 1971, he began writing an account of his life and the interesting people with whom he had been acquainted. He read extensively, but expressed the opinion that the events and characters of his early life were more interesting and colorful than most of the books he'd read! Indeed, a talented writer could construct many short stories, or even longer works, from the basic stories he told so often to his children and grandchildren.*[34]

Several of Wheeler's stories involve burros, which were frequently ornery and downright mean. Burros were used in the mines to haul equipment and ore but could also be found roaming free in Creede and Bachelor. As such, encounters between burros and curious children were common:

> *We tramped the hills, climbed the cliffs, went into old prospect holes, climbed down shafts and tried our darndest to end our lives, but somehow we survived. Burros were our main mode of transportation. The ornery, patient varmints. We were bucked off, rubbed off between trees or under clothes lines, stepped on, kicked, bitten, and rolled over on. They were as ornery as we were. We never rode with the saddle, always on the very rear deck with a rope or bailing wire for harness. I once rigged a rope around a burro's neck, then tied it to a board, on which my brother was supposed to ride. I was to lead the burro, but when he heard the board following him, he ran right over me. Broke my left collar bone.*[35]

Only boys with imagination and free time on their hands would work for days to experience a thirty-second thrill.[36] Because Bachelor was a short walk from mines on the Amethyst Vein, it was tempting to explore them in search of adventure:

> *Then there was the big project that took nearly all of one summer. The old Happy Thought shaft house had burned down. Out on the dump was the drum of the hoist. It must have weighed*

five or six tons. We got to wondering what would happen if it rolled on down the mountain, so we went to work. We hitched burros on to it and pried, but no move. Then we dug up under it. Finally, after moving tons of dump, it gave way. That was the thrill of a lifetime. That thing cut a swath through heavy timber on its way to the creek about half a mile below. Trees one foot and more in diameter were mowed down like match sticks.[37]

The remains of the Happy Thought Mine were the source of other adventures:

Also at the burned shaft house, there was a large tank, about twenty feet in diameter and about four feet deep. It had been used for water for the boiler. There was a small creek nearby, so we piped water to the tank. We had about 100 yards of pipe, tin trough, etc., which leaked plenty, but we did get a trickle of water to the tank. When the water got deep enough, we built fires around it. We had one dandy swimming hole. Peeled off our clothes and dived in. Bess, too. I think that was the first time I noticed that she was different.[38]

Edwin Lewis Bennett, whose family moved to the Creede area in 1893, relates stories of his youthful experiences at Bachelor in *Boom Town Boy*. His father worked at the Amethyst Mine, and they lived in Weaver, a small community just over the hill from Bachelor, before his family moved to Creede:

In March, 1906, Mr. E. R. Cantwell hired me to carry mail to Bachelor. How he got the contract, I'll never know, and why he got it is more of a mystery. He ran a harness shop on Second Street, alongside Del Wilson's store, but he was as little an outdoorsman as I was a harness maker. By the time he got through feeding a horse and paying me, there was little left from his contract pay, and he dropped it when it ran out. Six days a week, I started up the hill to Bachelor, two miles away on the top of Bachelor Mountain, making my start some thirty minutes after the morning train got in around ten o'clock and coming back down in time for the out-going mail to be sorted in time for the four-twenty departure of the train. My pay probably ran somewhere around one-fifty a day. Whatever it was, I turned it over to the folks, and made my spending money out of what I got for carrying small packages between the towns and from other activities.[39]

Another story took place while he was carrying mail to Bachelor:

Fred Foster, an old pal of mine, lived in Bachelor and his father owned a saloon there. At the time I was carrying mail, Fred was learning to mix drinks, and in me he found a willing guinea pig. From the Bartender's Handbook, or whatever they called it, he would mix up a drink, simple ones at first but, as he became more proficient, the complicated ones. The drink mixed, he poured it into two glasses and we passed judgement on it. If unsatisfactory, Fred figured maybe he had erred somewhere along the line and made another. If that was no better than the first, he went on to a new one . . . If the thing turned out to be something that appealed to us he mixed another to be sure he could do it again. Sometimes the second was not, in his judgement, up to the first one so he made up a third, but they all tasted so much alike to me I suspected he wanted that third one. We didn't go overboard on the project but, after four or five drinks, the mail carrier started down the hill somewhat jingled, which a mail carrier is not supposed to do. But a half hour in the open air cleared the fog and I was able to deliver the mail sack in good condition. These experiments continued for the whole time I carried the mail so, by the time I lost the job, both Fred and I were conditioned to where we could carry a respectable tonnage without outward showing.[40]

Because of this "trained" ability to handle alcohol, Bennett frequently drove people by wagon from Bachelor to Creede and back on dance night. He was chosen because he could hold his liquor better than the others, which was important for driving a team up and down the steep slope to Bachelor from Creede.[41]

Harold Wheeler describes how he and his friends welcomed a new teacher to Bachelor: "School had not opened yet, but the new man teacher was in town. We got up on top of a vacant store building with a false front and piled snow on top of it. When the new teacher came along the sidewalk below, we pushed off the snow. Nearly covered him up. He sure got even when school started. He brought along a raw hide whip. Whenever he used it, it left a red mark, and he used it every day."[42]

The teacher's name was Mr. Sail, and judging from the description of his accent, he must have been from the East Coast: "He said 'Octic' and 'Antoctic' circles. Until then I thought they were the Artic [sic] and Antartic [sic]. You learn something every day in school."[43] As the school year progressed, Mr. Sail continued to "instruct" the boys: "Archie and I were once caught throwing spit balls. He kept us after school and made us chew up a whole waste basket full of paper into spit balls, throw them and then clean up the mess. That just about put an end to the game."[44]

In an article by Jack Foster that appeared in the *Rocky Mountain News* in October 1952, he describes a trip he took to Bachelor. After breakfast at the Creede Hotel, his host, Orrin (Junior) Hargraves, suggested that they take Fred Ryden on the trip because Ryden had lived in Bachelor from 1893 to 1904. Ryden, who owned a drugstore in Creede, happily accepted the invitation and closed his store for the day.

Ryden shared stories from his childhood in Bachelor and described an experience he and a friend had one winter:

"One Christmas holiday," Fred remembered, "I left school with a friend, Steve Doering, to go by skis over the Continental Divide to the head of Spring Creek where his family had a ranch. He was 16, I was 15. And it took us two days. And a mountain lion followed us part of the way. But it was a wonderful holiday. How good that mountain sheep tasted for dinner!"

Upon hearing the story, Foster exclaimed, "Can you imagine a 15- and 16-year-old kid starting out today to cross the Continental Divide alone in dead of winter?"[45]

One final example of Harold Wheeler and his friends working for a long time for another thirty-second thrill involved the Wheeler home:

Rolling rocks off the cliff back of our house was a favorite sport. Rocks got so scarce that we practically had to pack them back up. There was one big rock up there that we couldn't move. After prying on it for two or three years, we finally got it started. Instead of rolling straight down the hill, like any respectable rock would, this one veered off and went through our kitchen. My father was in the house at the time, but happened to be in the bedroom. This episode was not so easily covered up. The rock is still there, in what was the back yard. I saw it two years ago when we were in Bachelor [during the summer of 1969].[46]

Culture

While Bachelor definitely had its rough side, and most people who lived there were hard workers, they also found time to enjoy musical and theatrical talents (see fig. 48).

Like other mining communities, there were parlor houses, gambling halls, and saloons. There must have been a little culture in Bachelor, however, for it had its own opera house, and the Bachelor City Dramatic Club which was touted as being excellent. The town also had its churches. The Congregational Church boasted that it was the highest church in the United States (the elevation of the town was 10,526 [feet above sea level]).[47]

Fig. 48. Photograph titled Bachelor Orchestra, 1893. It shows a man with a banjo and guitar, with his wife or fellow musician in the background. Creede Historical Society Archives.

One of the first tasks that Bachelor's residents undertook was seeking funding for a new Catholic Church (see Chapter 5). A fundraising event hosted by the Ladies Aid Society was held on August 16, 1892, and the *Teller Topics* described the event in great detail, with tones that sound like something out of the society pages of a big-city newspaper:

Last Tuesday evening will long be remembered in Bachelor as the first social event of the season—worthy of the cause for which it was given, to aid the building of a Catholic church in our city. The hall was tastefully decorated and everything passed off in good shape. Much credit is due to the reception and floor committees and to the ladies who so ably managed the affair. Financially it was a success, as fully $250 was the result of sale of tickets. The programme of the evening, which in itself shows get up, we publish in full:

PROGRAMME

1. Grand March .... "Our Miners"
2. Quadrille........................ "Free Coinage"
3. Waltz............................ "Amethyst"
4. Lancers.......................... "Last Chance"
5. Schottische..................... "Del Monte"
6. Quadrille....................... "Hidden Treasure"
7. Newport........................ "Bachelor"
8. Portland Fancy................. "Our City Fathers"
9. Waltz........................... "Allanby"
10. Quadrille...................... "Capt. Campbell"
11. Virginia Reel.................. "Crawford"
12. Polka.......................... "Eight Hours"
13. Quadrille...................... "Our Visiting Friend'
14. Waltz.......................... "Silver Dollar"
15. Lancers........................ "General Weaver"
16. Quadrille...................... "E. Downey"
17. Schottische.................... "Jimtown"
18. Cicilian Circle................ "Creede"
19. Waltz.......................... "Wagon Wheel Gap"
20. Lancers........................ "Lime Creek"
21. Polka.......................... "Aspen"
22. Waltz Quadrille................ "Leadville"
23. Mocking Bird Quadrille...... "Burro"
24. Waltz.......................... "Good Morning", "Home Sweet Home."

It was early dawn before the social party separated, pleased beyond measure with the evening's entertainment.[48]

Because it advertised, promoted, and covered such events, *Teller Topics* itself was part of the culture of the town. Its presence was a magnet for the growth of culture and business. The paper addressed this role in its September 3, 1892, issue:

Since The Topics *began publication, Bachelor has grown from a collection of miners' cabins into a prosperous business center and is on the road to become the metropolis of Creede camp. We claim a goodly share of credit and propose to go on advocating the superior attractions of the town and drawing capital to it. The paper is past the crucial state and has become a permanent feature of the camp. We extend thanks to the business men who have made these successes possible and hope to continue to merit their support.*[49]

Indeed, its pages embodied the culture and spirit of the young town.

Because of the close proximity of the mines to the town and the fact that their destinies were so intertwined, many of the social and cultural events were a joint effort of the Last Chance Mine and Bachelor. One such event was the opening of the new boardinghouse at the Last Chance Mine in the summer of 1892. The *Teller Topics* describes the elaborate event as "a Social and Artistic Success," continuing:

Last night the social event of the season occurred. It was the opening of the spacious new hotel which the Last Chance company has built at the mine for the accommodation of such of its employees as may wish to avail themselves of it.

The large dining hall was tastily decorated and nothing that promised to contribute to the enjoyment of the occasion was left undone. The miners, the merchants, the business and professional men from all over the camp came with their wives and sweethearts, filling the building and making as social a party as it has ever been our privilege to mingle with.

Excellent music and prompting was at hand and the evening passed swiftly by. Fifty couples occupied the floor continually. The programme of dances was happily arranged to make the evening thoroughly enjoyable. Under the able direction of Floor Manager Marcell the poetry of motion was illustrated to perfection.

Promptly at 12 o'clock the signal for refreshments was given and a dozen young ladies dressed in pretty Turkish costumes, filed in from the kitchen, each bearing a salver heaped with generous packages of edibles concealed in tasty Japanese papers carrying the compliments of the Last Chance. Making a graceful salaam, the waitresses presented to each guest a package which, on being opened, was found to contain an abundance of fruits, cakes and bon bons....

Another waltz, a quadrille and with "Home Sweet Home" the long remembered occasion was over.[50]

Bachelor had its own Dramatic Club that performed in the Bachelor Opera House. As reported in the *Creede Candle*, the group performed the *Wild Irishman* in December 1892 at another successful fund-raising event for the new Catholic Church:

Bachelor people outdid their previous record for good entertainments in the one given Tuesday night for the benefit of Father Downey. The new opera house was packed and everybody declared that it was the finest affair ever given in the town.

The Bachelor City Dramatic Club furnished the programme. The drama of the 'Wild Irishman' was given. The acting was declared to be much above the usual amateur attempts, and the various actors displayed good talent in their several lines.

Messrs. Walker and Bishop and Mesdames Nash and Marsell gave a vocal quartette as the opening number.

Then followed the drama with the following caste [sic]:

Delaney . . . . F. Bishop
Aunt Prudence . . . . . . Mrs. Collins
Mrs. Bella . . . . Mrs. A. McDonald
Eugene Fitzgibbon . . . H. Welch
Rudolph St. Clair . . . . W. Laird
Jiles . . . . J. Gardner
Podkins . . . . Mr. Marsell
Peggy . . . . Mrs. J. Leonard

J. Gardner recited 'Shamus O'Brien.' Walker and Bishop gave a duette [sic]. C. J. Gavin recited 'St. Peter at the Gate.' Messrs. O'Leary & McWade gave a double clog act which brought down the house. Miss Born and Mrs. Marsell sang a duett [sic] in a highly pleasing manner.

Encores were frequent and enthusiastic.

A dance followed and was heartily enjoyed. About $160 was realized for the purpose given.

J. W. Winthrow contributed largely to the success of the entertainment by his direction and advice on delivery of parts.[51]

There were several excellent photographers in the Creede Mining District. Two of the best—Charles Goodman and E. F. Hilton—had studios in Bachelor. In many ways, photographers were the artists of the time because they captured on film many of the views from the area. An example is the panoramic view shown in Figure 1. the *Teller Topics* said much the same. In the August 13 issue, the newspaper reported:

An old adage: 'The poet is born, not made.' This applies equally to the artist. Those of our readers who have not examined the scenic work of Charles Goodman, offering views of Batchelor [sic], the tramway and other points, would do well to call at Ellsberry's furnishing store and take a look. For good scenic and well-defined perspective, the views referred to, show the talent of the artist to perfection.[52]

Examples of Goodman's photography are seen in Figures 31 and 40.

A steady beat of social events in 1892 was reported in the *Teller Topics*. On August 6 the paper reported: "Quite a sociable party was given by Miss Agnes Stone yesterday evening at the Commercial House . . . An elegant supper was served and all enjoyed themselves to their hearts content."[53]

A very pleasant party was given at the Teller house on the evening of the 4th in honor of the birthday of Master [sic] Harry Whitehead and Scott Matlock. Good music was listened to and refreshments were served by Mrs. Whitehead, who entertained her young guests in royal style.[54]

In the same issue, the newspaper reported that Professor T. F. Leary was opening a dancing academy on Park Street[55] and stated: "James Emerson of the Theatre Comique, Jimtown, was in the city Thursday making arrangements to give a show twice a week at Hubbard's hall."[56]

July 4, 1894 Celebration

Fourth of July celebrations were major events in western mining camps in the late 1800s and early 1900s. They included parades, foot and horse races, and various mining competitions. Many of the towns carry on this tradition to this day—Creede is an example.

The Fourth of July celebration in Bachelor in 1894 was special because the town had helped Creede with its 1893 celebration. Creede was honor-bound to reciprocate and the June 29, 1894, issue of *Creede Candle* reported that:

> *The following resolution was passed by the Creede Board of Trade at a recent meeting. The advice is good and should be followed:*
>
> *Whereas, it is the apparent desire of the people of Bachelor to make a success of their celebration and exercises on July 4th,*
>
> *Be it resolved, That inasmuch as they, by their presence and encouragement materially aided our city in its celebration last year, the Creede Board of Trade heartily endorses their efforts and earnestly requests the people of our city to join with the citizens of Bachelor in the efforts to make it a day memorable in the history of our camp.*[57]

In the same issue, the *Creede Candle* urged Creede residents to go to Bachelor and help with the celebration.[58]

A grand ball was planned at the Last Chance Mine boardinghouse on the evening of July 3. Tickets for a lady and gentleman, including dance and supper, cost $1.50. The Barry and Zumiebel Orchestra was slated to play at the ball on the evening of July 3 and at the Bachelor Opera House on July 4.[59]

The *Creede Candle* listed the events and prizes for the big day. Arrangements were also made for people to bet on the events:

PROGRAM:

Exercises commence at 1 o'clock with music by the band.

1. Girls' burro race, 4 to enter, prize, $5.

2. Boys' sack race, 4 to enter, prize, $5

3. Boys' foot race, 4 to enter, under 15 years,
   prize $5.

4. Drilling contest, 15 minutes to team, 2:30
   o'clock, 1st prize, $250; 2nd prize, $150.
   3rd prize, $50. Judges—E. H. Crawford, J. Allenby and
   Walter Boyle. Time-keepers—J. Rodman and Capt. L. E.
   Campbell.
   Entries to be made before July 1; $10 per
   team. Rules governing contest can be seen
   at Gem saloon.

5. Climbing greased pole, prize, $10 bill on
   top of pole.

6. Grand Miners' ball at Vincent's new opera
   house; four pieces of music.

Two handsome fancy silk vests have been donated
by J. McKenzie, the Bachelor tailor, for the winners
of the first prize in the drilling contest.

The Creede band has been engaged to furnish
 music during the festivities.

Books will be made and pools sold on the drilling
contest at the Gem saloon.

EVERYBODY TURN OUT and have a rousing
good time.[60]

The miners' drilling contest was the most anticipated event. It also paid the largest prizes, including $250 for the winners. This was equivalent to nearly six-months' pay for a miner, so interest in the contest was high. An early indication that there would be a good number of contestants was the large demand for hammers and hammer handles at Tomkins Bros. store: "It is interesting to notice the care exercised by the miners in the selection of their practice tools and the pride the [sic] feel in having nothing but the finest and best of implements."[61]

On Tuesday, June 26, the block of granite to be used in the contest arrived at Creede on a Denver & Rio Grande Railroad flat car. While a 3,000 pound rock was ordered, it was found to be closer to 7,000 pounds.

The large stone was hauled up to Bachelor the next day on a wagon pulled by a team of eight head of stock (probably oxen) driven by Joe Jenkins:

The granite is very handsome and while good and hard it is predicted that some fine work will be done in the drilling contest as it contains no fissures. The block comes from the stone quarries just south of Gunnison City, the same that the material in the state capitol at Denver was taken from.[62]

In the same article it was noted that James H. Collins, a Creede businessman who once owned the quarry, remarked as he examined the block of granite that he had not had the good sense to keep it.

The celebrants on July 3 and July 4 had a great time. Reports were that everybody went to Bachelor to celebrate the birthday of the United States and that the day went as planned. The burro races and other events provided the expected amusement, but the main interest centered on the double drilling match:

Money changed rapidly over the contest. The teams had practiced considerably and many felt able to pick the winners but the result was a surprise party all around. Following is the score:

DRILLING CONTEST.
Judges – E. H. Crawford, W. I. Covert and Walter Boyle.
Timekeepers – J. Rodman, L. C. Hunt and J. D. Jennings.
1st Team – Will Crawford and Ike May. 2.12 o'clock, 26 1/2 inches.
2nd Team – Jack Lamb and Matt McCarty. 2.37 o'clock, 24 1/2 inches.
3rd Team – Harry Welch and Bill Lamb. 3.06 o'clock, 28 1/8 inches.
4th Team – Jack Regan and Harry McMullin. 3.28 o'clock, 30 3/4 inches.
5th Team – Colin McGilvray and James Driver. 3:15 o'clock 30 1/8 inches.

"HOBO" Drilling Contest.
Del Wilson and Charles Johnson, 19 7/8 inches in 10 minutes.[63]

The second article on page one of the July 6 *Creede Candle* gave a running summary of everything that happened, captured in an almost poetic style. Whether by design or coincidence, most lines were progressively longer.

Everybody was 'coaching.'
Most of the band walked up.
Everybody yelled and danced.
A good many are poorer but wiser.
Only one drill stuck on the winning team.
'Pretty work.' And it was, indeed
Regan and McMullin were a 'dark' team.
What a team Driver and Regan would make.
The noisy and talkative man was in his glory.
Chas. Duncan and his wife were over from Weaver.
The state record was smashed all to smithereens.
Six teams contested in the 'Hobo' drilling contest.
Some of the 'coachers' worked as hard as the teams did.
Some of the teams were not evenly enough matched.
Old Man Winn did not win that $10 bill. Too much grease.
McGilvray was absolutely blue in the face when 'time' was called.
Joe Jenkins worked hard and was here, there and everywhere.
The 'milk shake' and 'circus lemonade' stands did a good business.
County Clerk Mark G. Woodruff was among the outside timekeepers.
'Alabam' was strictly 'in it' and coached hard for the sawmill team.
County Commissioner Frank E. Wheeler was at the ropes watch in hand.
The livery stables were cleaned out of everything that could furnish transportation.
Mrs. Reidmiller opened a branch of her business in Bachelor during the celebration.
Welch and Lamb made good changes and their drills stood well. They took third money.
The grand stand was crowded but was too far away for people to see much of what was going on.
The man with the striking machine had a little mint of his own for a few hours and just 'coined money.'

Wilson and Johnson, winners of the 'hobo' drilling contest, would have given some of the winners a close race.

Bets were as freely made and as freely taken that the Welch and Lamb record, 28-1/8 inches would not be beaten.

Harry McCulloch sharpened and tempered the drills for the 'hobo' contest. Out of the 60 drills not one of them broke.

The 'sawmill team' had 30 inches drilled at 11 minutes and a bet of $100 was made that they would win 1st prize.

Reagan and McMullin threw their drills so hard in making changes that the spectators had to 'look them a leedle oud.'

The team that sunk the deepest hole while practicing, Lamb and Welch, only got third place, while the second best team at practice, Lamb and McCarty, brought up the rear in the contest.

Some Creede and Bachelor business men are talking of forming a team composed of Driver and Reagan, making up a $500 purse and challenging Leadville to a rock drilling contest. A good idea.

Driver and McGilvray made good changes and did great work. Had their last two drills not stuck they would probably have taken first money and would have given the leaders a close rub anyway had Driver struck during the last minute.[64]

A quarter-mile horse race was won by James Workman's Minerva. Henry Atkinson's Crooked John finished second. The Workman horse was ridden by Cleve Braley and Atkinson rode Crooked John. The race was close from start to finish, with Minerva winning by a head: "About 200 people were present among whom there were quite a number of ladies and everyone seemed to have a good time."[65]

So, the festivities of July 3 and July 4, 1894, had ended, living up to high expectations. As people wandered home in the early hours of July 5, they were confident that they had celebrated the anniversary of the great country in "glorious" fashion.

Chapter 7
A Town in Decline

Silver Politics

The "silver boom" in Bachelor, Creede, and other western silver mining camps was dependent on the purchase of silver by the U. S. government. This, of course, was contrary to the law of supply and demand and placed the camps in the tenuous position of depending on continued government purchases. One of the major factors leading to the decline of western silver mining towns in the 1890s was the end of government support of silver.

The U. S. currency system was originally founded on bimetallism, a concept that required the government to produce both gold and silver coins. The ratio in value was set at sixteen to one, which meant the value of gold in a gold dollar was sixteen times that of silver in a silver dollar. Unfortunately, this arbitrary ratio did not take market factors into account.

When gold was discovered in California in 1849, the increased supply of gold dropped its price per ounce. At the time, silver was in shorter supply, so its price rose as a result of the law of supply and demand. In 1868, for example, the price of silver was so high that it was profitable to melt silver coins and sell them as bullion. As a result, the U.S. Congress passed the Coinage Act of 1873 that ended the coinage of silver dollars and restored the law of supply and demand. It also established the gold standard, which made sense because the price of gold was more stable than that of silver. Silver advocates were outraged and branded this act "the crime of '73."[1]

When silver was discovered in western states in the 1870s and 1880s, the price of silver per ounce dropped as a result of the increased supply. This led silver-producing interests and western states to demand that the government return to bimetallism. Their cause was joined by members of Congress who wanted cheap money to stimulate the economy out of a five-year recession. As a result, the Bland-Allison Act of 1878 was passed over President Hayes's veto. It required the U. S. Treasury to buy 2 million to 4 million dollars in silver each month from western mines and convert the metal into silver dollars.[2]

The argument over gold and silver continued and became a focus of tension between western states, which favored free coinage of silver regardless of the quantity produced, and eastern states, which favored a single gold standard. Because Idaho, Montana, Washington, Wyoming, South Dakota, and North Dakota had recently achieved statehood, the western states had enough power in Congress to pass the Sherman Silver Purchase Act of 1890. The new law did not authorize the free coinage of silver that silver advocates wanted, but it did require the U. S. Treasury to purchase an additional 4.5 million ounces of silver each month. The law required the Treasury to buy the silver with a special issue of treasury notes that could be redeemed for either silver or gold.[3]

The price of silver quickly increased from $0.84 to $1.50 per troy ounce following passage of the act, fueling the start of the boom that took place in Creede and Bachelor in the early 1890s. However, the high prices did not hold. In March 1892, the price had fallen to $0.85 an ounce.[4]

Passage of the Sherman Act had another negative effect that ultimately led to its repeal. While it did increase the circulation of redeemable paper currency in the form of treasury notes, it also created a drain on U. S. gold reserves because investors began redeeming the new treasury notes for gold rather than silver.[5] The issue came to a head in 1893 when the Panic of 1893 occurred, partly in response to the U. S. equivocation on the gold standard. In response, President Grover Cleveland took an active role in seeking repeal of the Sherman Silver Purchase Act to prevent the depletion of U. S. gold reserves. He called Congress into a special session on August 7, 1893, to deal with the issue. After considerable debate, particularly in the Senate, the Sherman Act was repealed in October, with a House vote of 239–108 and a Senate vote of 48–37.[6]

Repeal of the Sherman Act had a chilling effect on silver mining throughout the western states, and "officially" ended the booms in Bachelor, Creede, and other western silver mining towns. The price of silver dropped from $0.83 to $0.62 per troy ounce in the four days after repeal. While repeal did not shut down mining on Bachelor Mountain, it had a severe effect.[7]

Cuts in mining staffs began even before repeal of the Sherman Act because of the drop in silver prices. The June 30, 1893, issue of the *Denver Republican* reported:

All the producing camps have practically stopped shipping, the owners preferring to let the silver lie in the ground rather than sell it at the present low price. The Amethyst was the first to reduce force and did so early in the week. The New York and Chance and Bachelor will ship no more after tomorrow until the price goes up. Development is going ahead in all the big mines without intermission, and all are working small forces to keep out the water, protect the properties, and open new ground.[8]

An article in the July 7, 1893, issue of the *Creede Candle* reported layoffs at the Last Chance Mine:

The Last Chance which had been running full-handed up to that time, cut its force in half yesterday, letting out seventy-five men and continuing that number at work. They were the last to close for the reason that their contract with the Leadville smelters kept them up. The New York and Chance is completely shut down but the Amethyst continues to give employment to about 40 in developing and protecting the property.

Development work goes on all over the camp with little change because of the depression in the silver market.[9]

While employment at the mines recovered somewhat and production resumed, the peak had been reached and things would never be the same again. Realizing this, people began to leave Bachelor. Some moved to Creede, but many left the area entirely looking for better prospects. Among the first to leave were the gamblers, prostitutes, con men, and others who depended on "boom times" to make a living.

Impact of Construction of Mining Tunnels

As mentioned previously, Bachelor's main reason for existence was its close proximity to the major mines on the Amethyst Vein. At the time the town was formed, the mines could only be accessed from the surface, so miners living in Bachelor had a short walk to the mine shafts. The alternative access from Creede was up West Willow Canyon followed by a steep climb up to the mines or a long ride from Creede to Bachelor and on to the mines. It is not surprising that people preferred to live close to the mines. This changed when tunnels were dug that allowed miners to access the mines from the bottom. It also created a major advantage for the mines, allowing them to bring out ore through the tunnels rather than raising it to the surface.

What began as the Nelson Tunnel was the brainchild of Charles F. Nelson. He saw it as an opportunity to drain the mines of water and bring ore out through the tunnel rather than by the expensive process of raising ore to the surface. When completed, it was one of Colorado's greatest tunnel achievements.[10]

The Nelson Tunnel was driven 2,100 feet northwest from the portal (fig. 49). The Wooster Tunnel branched from the Nelson Tunnel at the 1,400-foot point and followed the Amethyst Vein for 6,900 feet to the Amethyst Mine. The Humphreys Tunnel was a 3,900-foot extension of the Wooster Tunnel to the Happy Thought and Park Regent Mines: "The tunnel intersected the Last Chance, Amethyst, Happy Thought, and Park Regent shafts at a depth of 1,500 below the apex of the Amethyst Vein."[11]

Because the Nelson Tunnel failed to meet its objective to intersect the Amethyst Vein, operations ceased and the company went bankrupt. In 1897 Albert E. Humphreys, owner of the United Mines properties on the Amethyst Vein, helped form the Wooster Tunnel Company and took over the Nelson Tunnel under a ninety-nine-year lease. As mentioned, the new tunnel took off from the Nelson Tunnel at the 1,400-foot point and encountered the Amethyst Vein after 400 feet of drilling. Over the next several years, the tunnel advanced along the vein until 1899, when the Amethyst and Last Chance Mines were encountered and drained:

By March 1902, the Wooster Tunnel Company had contracts with the Bachelor, Amethyst, Last Chance, New York, and Chance Mines, and United Mines to transport all their ore below the 400-foot levels through the tunnel. It was charging an average of $1 per ton. Ore was being delivered to the new Humphreys Mill over a track extended about 2,400 feet beyond the portal as a surface tramway to the mill, equipped with mule haulage.[12]

Fig. 49. Entrance to Nelson Tunnel, ca. 1892. Following bankruptcy and a change of ownership, it ultimately became the Nelson-Wooster-Humphreys Tunnel. Creede Historical Society Archives.

When completed, the Nelson-Wooster-Humphreys Tunnel provided a less expensive outlet for ore that could be directly conveyed to the Humphreys Mill.[13] Once processed, it was loaded directly onto Denver and Rio Grande Railroad ore cars for transportation to various smelters. The tunnel eliminated the need for the machinery to lift ore and pump water from the mines. Instead, ore and water could be brought out the bottom of the mines through the tunnel. By draining water, the tunnel also opened up new ore bodies. Another major benefit was that the tunnel greatly reduced the need for overland wagons to haul ore from the surface of the mines to ore dumps or train cars. It also allowed mines to resupply through the tunnel rather than surface facilities. Most important, it largely eliminated the need for the town of Bachelor, which depended on activity to and from the surface of the mines for its economic health.

Edwin Lewis Bennett, who moved with his family to Weaver in 1893, describes the impact of the tunnel on the towns on Bachelor Mountain:

When we lived in Weaver, the town of Bachelor stretched all the way over the top of the hill and included a string of cabins above the top workings of the mines. Because the cabins were scattered out among the old stumps left by the timber-cutters when they were getting out mine-timbers, it went by the name of Stumptown, although for voting and taxing purposes it probably was a part of Bachelor. Between that period and the time I carried the mail, the Nelson Tunnel had been driven and the miners no longer went into the mines through the top workings but through the tunnel, and that took away Stumptown's reason for being so that Bachelor ended far short of its old boundaries. The same reason had caused many Bachelor people to move down to Creede and started a decline that ended only when the last settler left.[14]

Decline in People and Businesses: 1892–1910

No good information is available on the population of Bachelor in the years following the collapse of the price of silver. As mentioned, the population was around 350 (or 700) in March 1892 (see fig. 20) and peaked in the range of 1200–1500. The next official population reading was from the 1900 U. S. Census, which listed 334 Bachelor residents (fig. 50). The rate of population loss between 1893 and 1900 is unknown, but the number of business listings, also shown in figure 50, remained fairly stable until 1896. This is supported by a photograph of the town in the summer of 1895 (see fig. 1) that appears to show it was still thriving. Again, based on business listings, there was a steep drop from 1896 to 1897. This is coincident with progress on the Nelson-Wooster-Humphreys Tunnel and suggests that, while the town survived the initial blow from the 1893 repeal of the Sherman Silver Purchase Act, it finally succumbed to the impact of the tunnel.

Population and Business Trends in Bachelor

Year	Population	Business Listings[1]
1892	ca. 350 (or 700)[2]	0
1893	1200–1500	60
1894	800[4]	58
1895	800[4]	60
1896	800[4]	51
1897	300[4]	28
1898	200[4]	14
1899	200[4]	15
1900	334[5]	15
1910	180[5]	2
1915	na	0

[1] Taken from introductions in Colorado Business Directories, various years.[15]
[2] Informal census taken for town incorporation. See Figure 20.
[3] See population discussion in chapter 3. Colorado Business Directory listed 6,000.
[4] Taken from the *Colorado Business Directory* listing for each year.
[5] Taken from U. S. Census data at the Creede Historical Society Library.

Fig. 50. Population and business statistics from 1892 through 1915. The decline in population and businesses became pronounced in 1897. Despite this, life in Bachelor remained fairly normal until the early 1900s.

A harbinger of things to come occurred in July 1893, when the Ellsberry & Foutch Banking and Mercantile Company failed (see Chapter 5). This was the only bank in Bachelor and dealt a severe blow to the town.

The school in Bachelor continued to operate, although at a considerably reduced level. A 1910 front-page article in the *Creede Candle* illustrates the point. It reported scholars at Teller School had presented a lengthy program that included recitations, readings, a piano solo, and a choral presentation. The article concluded with this account of enrollment at the school:

Enrollment 29, 15 boys and 14 girls. Per Centage [sic] of attendance, 95. School term 7 months. Number in classes as follows: Beginning class, 5; First grade, 2; Second grade, 3; Third grade, 9; Sixth grade, 1; Seventh grade, 4; Eighth grade 4; Ninth grade, 1. Graduates from the Eighth grade, Oscar David, Burnald Dean. Sarah E. Reath, Teacher."

The low enrollment was typical for a small, rural school, but it also indicated that the school was on the edge of viability.

Things became more difficult the next school year. The term began with twenty-four students, but by the end of the year enrollment had dropped to eight. Expulsions were given as the reason for the large attrition.[17]

Telephone Service

Telephone service did not arrive in the Creede/Bachelor area until 1906. Regular phone service began in Creede in early January, and the telephone company treated residents to a free phone day to celebrate the event.[18] Telephone lines were extended as far as Upper Creede in February, but work was suspended because conditions became too difficult in the canyon:

The reason for this sudden termination of construction work is owing to the enormous expense they are under while trying to work in the frozen ground, deep snow and cold weather in the canon. Foreman Bliss informs us that it has cost as high as $25 to set one pole under present conditions and they are absolutely unable to accomplish any headway.[19]

Temporary connections were made to Bachelor and surroundings pending resumption of work when weather conditions permitted. Work resumed in June 1906.

The telephone construction crew arrived the first of the week and have been busy completing the line work left undone last winter and putting in the new line to the mines and Bachelor. They are also painting the poles. There are seven or eight men in the gang at present which will be considerably increased as men can be spared from other districts."[20]

A report in an August issue of the *Creede Candle* stated that the work was "pretty near completed."[21] Bachelor had permanent phone service (see figs. 51 and 52).

Fig. 51. Postcard view of Main Street (cap. 1910) showing telephone poles lining the east side of the street. Harbert Archives.

Fig. 52. View of the business section on Main Street, ca. 1910, looking downhill. Telephone poles extend down the left (east) side. Note that sidewalks extend across the street as well as in front of the businesses. Harbert Archives.

Fires Still a Present Danger

Fires continued to be a problem in Bachelor as they had from the beginning. Considering that many homes and businesses were heated with wood stoves and lighted with candles and lanterns, it was a surprise that there weren't more fires. In 1905 a fire struck the home of Professor Kent, a teacher at the Bachelor Public School.

> *Prof. A. R. Kent's home in Bachelor was burned to the ground Wednesday morning with all its contents. The fire started from a defective flue in the kitchen and before discovered it was too late to even save the occupants' wearing apparel. Mr. Kent was at school and before he or assistance could reach the building it was entirely enveloped in flames, making it impossible to rescue anythidg [sic] and all that the family have [sic] left is what was on their backs. Mr. and Mrs. Kent and son departed Thursday evening for a couple of days visit to Demver [sic].*[22]

Another disastrous fire occurred on Thanksgiving Day 1911:

> *Thanksgiving Day in Bachelor was not without its excitement, albeit the excitement did not really occur until about eight o'clock at night.*
>
> *To quote the immortal 'Scotty' Berry: 'Thanksgiving night in Bachelor is just the same as New Year's night in New York City, or Twelfth night in Glasgow.'*
>
> *Things were strumming along very nicely, so says Mr. Berry, when someone came out of their trance and started the fun by setting a building on fire, through the medium of an overheated stovepipe. The fact that Bachelor has practically no fire fighting apparatus beyond a bucket brigade caused at least three buildings to fall a [sic] prey to the flames.*
>
> *The structures destroyed were located between the store operated by A. B. Eades and Mr. Berry's residence. It is estimated that the loss will be in the neighborhood of $1,200.*[23]

Another serious fire was narrowly averted at the George Davis saloon. The blaze started on the second floor and was burning fiercely when discovered. Hose Company No. 2 rushed to the scene and used a bucket brigade to put out the fire. The upper rooms were badly scorched, but there was little damage otherwise. As with the other fires, the cause was a defective flue.[24]

Automobiling to Bachelor

With the advent of the automobile, the steep dirt road from Creede to Bachelor became a challenge that avid motorists could not pass up (fig. 53). At the time, the distance from Creede to Bachelor was about 2.5 miles (see fig. 11) on a road that was considerably shorter and steeper than the current Forestry Service Road 504, which is 3.7 miles with a moderate grade. It also became a sport for the people of Creede to watch the ascent because much of the road was visible from the town.

Fig. 53. Photograph of four horse riders on the road from Creede to Bachelor (ca. 1900). The view shows the condition of the road at the time. It was probably worse when the first car navigated it in 1907. Creede Historical Society Archives.

Cyrus Miller was apparently the first to successfully make the climb in 1907. Miller, superintendent of the Amethyst Mine, drove his Stanley Steamer automobile from Creede through Bachelor to the mine. In 1910 two attempts were made by a guest and a member of the Humphreys family, who owned a lodge near Wagon Wheel Gap:

If you hear of any automobile manufacturer or agency wishing to demonstrate the hill climbing capabilities of his particular machine send them on. We have the mountain and the road and the pace has been set. We refer to Bachelor Mountain and the road leading to the town of that name, Bachelor, elevation 10,600 feet at the town pump. The elevation at Creede is 8,850 feet and the distance to Bachelor is called three miles but said to be a little short of that.

On Sunday last, Mr. Mat. Foster of the Humphrey's [sic] party drove to Creede from Wagon Wheel Gap and tackled Bachelor Mountain in his Packard car. Not anticipating the trip when leaving the Gap, Mr. Foster found a shortage of gasoline when about half way up the mountain and was compelled to return. He had demonstrated, however, that he could make the grade and on Monday returned with [a] full tank of gasoline and made the trip from Creede to Bachelor without diffult [sic], the only danger showing up being his fear that he might not be able to hold the car returning down the mountain to Creede but this descent was accomplished without accident.

On Tuesday morning Mr. Ira B. Humphreys drove up from the Gap and with his Apperson Jack-Rabbit car made on [sic] attack on old Bachelor Mountain with the same result as Mr. Foster had in his first attempt. When more than half way up Mr. Humphreys found a shortage of gasoline and returned to Creede but states that he will again make the trip and see to it that he has sufficient fluid to go through.

These events have caused considerable rubber-necking by our people as the Bachelor road is in full view from town most of the way and while the boys gained the advantage of the housetops the girls climbed the trees to see the fun.

Three years ago this season, the late Cyrus Miller made the climb from Creede to Bachelor and continued down on the other side to the Amethyst mine but stated that the experiment cost him a full set of tires for his Stanley Steamer machine. Mr. Miller made the trip immediately after arriving in Creede on a run from Denver.

No other attempt has been made since that time until these of the Humphrey [sic] party.

The road to Bachelor is in good condition and the grade averages twelve percent while in several places it more than doubles this.

We would like to see Mrs. Genevieve Phipps try this climb with her handsome Italian car which has been on our streets frequently this week.[25]

Mr. Humphreys was more successful in his second attempt:

Thursday morning Ira B. Humphreys drove up from Wagon Wheel Gap with his wife in his Apperson car and stopped in front of the Tomkins Hardware Company's store where he apparently replenished his supply of gasoline. He turned and made directly for the Bachelor hill, skimming up the road to the top of the mesa in an easy manner. We climbed to the top of the Collins Block and lost sight of the car just at the bottom of the steep red pitch. We were told afterward that he waited for a team to get down the hill and finally reached Bachelor, leaving there about noon. Ira feels considerably elated over his victory and carries a card bearing the stamp of the Teller post-office. What's the matter with the $12,000 Italian car we have heard so much about.[26]

Mrs. Genevieve Phipps, mentioned in the first article, was the wife of Colorado Senator Lawrence Phipps, owner of La Garita Ranch located about five miles south of Creede. She owned the Italian car mentioned in both articles.

Another automobile trip to Bachelor was reported in 1916, this time on a road that was apparently in worse condition than it had been when the earlier climbs were made:

Sunday afternoon A. F. Cooley and party of friends, from Del Norte, motored to Creede and after visiting the North town, decided to see the old town of Bachelor. We are informed that the trip was made without difficulty. The only stops were to throw out rock too big for the car to pass over. The trip was made in a seven passenger Studebaker 'Six' with a load of passengers. We do not know whether Mr. Cooley had been misinformed as to the condition of the road, or else, he wanted to know if his car could make the climb.[27]

It is interesting that the newspaper was already calling Bachelor the "old town."

Bachelor Voting Precinct Discontinued

Bachelor voters were located in voting precinct No. 7 for many years following the boom days. However, in October 1914 a notice to voters announced "Notice is hereby given to the voters of what was formerly known as Bachelor Precinct No. 7 [that the precinct] was discontinued by order of the Board of County Commissioners of Mineral County, Colo., October 2, 1914."[28] Voters in Precinct 7 were asked to appear at the Office of the County Clerk to be reassigned to Precincts 6 or 3, depending on where they lived.

Bachelor Congregational Church Torn Down

A major blow to Bachelor's dignity was reported in 1913 in the *Creede Candle*: "W. C. Duncan is busily engaged in tearing down the structure of the Congregational church in Bachelor, intending to use the lumber for improvements upon his residence here."[29] This raised the ire of Scotty Berry, a well-known Bachelor resident, who wrote a letter to the editor of the newspaper:

> *Dear Sir:—The Bachelor Congregational church that has stood so long in our midst, representing Christianity, went to the guillontine [sic] this week, presumably paying mammon its usual toll and giving to Shylock the usual pound of flesh. "Scotty" Berry.*[30]

People Moving from Bachelor

In her book *Stampede to Timberline*, Muriel Sibell Wolle discusses Bachelor's decline: "Bachelor flourished between 1892 and 1908 . . . But gradually the people moved down the hill to Jimtown below Creede, and the camp was deserted."[31] Wolle's timeline seems reasonable, but people actually started leaving Bachelor in the mid-1890s, and the pace picked up dramatically in the late 1890s and early 1900s. Articles about people leaving Bachelor began to appear regularly in the *Creede Candle*. It was a sad time in Bachelor, for both those leaving and those staying for a while longer. Many people relocated to Creede or Jimtown, while others left the area entirely—either for other parts of Colorado or out of the state. Several notable examples are discussed below, but others were not cited for the sake of brevity.

Tom Vincent sold his business interests in Bachelor to Pat Lundy in 1906.[32] This is significant because Vincent was one of the founders of the town as well as a well-respected businessman and citizen. He served on the first Board of Directors, was the first town recorder, and later became a justice of the peace. It is not known when he moved or to where, but he is not listed in the 1910 US Census of Bachelor. He was the last, or one of the last, of the town founders to leave.

Another big loss occurred in 1906 when Will Wood, Bachelor's postmaster and a well-respected businessman, and his family moved to Paonia, Colorado: "Will Woods [sic] has sold his entire stock of merchandise in Bachelor and is preparing to move his family to Paonia where he has a well established business. . . . Mr. Woods [sic] has been in business in our district for almost ten years and the many friends he has contracted during that time regret very much his departure."[33] Ironically, he also sold his merchandise to Pat Lundy.

Mike Fleming bought a piece of ground in the lower end of Creede in 1906 with the plan to move his buildings from Bachelor to Creede.[34] The next year he is quoted as saying he planned to move because he didn't like the winters in Bachelor, but he commented that Bachelor was a livelier town than Creede—even in 1907.[35] Ed Shelheimer, a Bachelor saloon owner, and his family moved to Creede in 1906.[36] Bill Jordan moved his family to Creede in August 1908.[37]

Fig. 54. Postcard of Main Street looking north, ca. 1915. The town is largely intact, but several buildings have been removed on the left (west) side of the street. Some were probably torn down and reassembled in Creede. Harbert Archives

In 1908, the *Creede Candle* reported what was already known: "Quite a number of families are moving down from Bachelor, where they will get more enjoyment in this life. We welcome them."[38] Visual evidence of the exodus is evident from photographs taken between 1910 and 1920, which show fences still present, but the main buildings were gone (fig. 54).

The Creede Historical Society owns a cabin that was moved from Bachelor to Creede. It is located on the north side of its museum. The cabin was restored in 2014 with the intention of opening it as an exhibit. The logs on each side of the cabin are etched with a number and a direction, indicating that it was disassembled and reassembled. The numbers on the northeast corner of the building are the most legible (fig. 55).

Some Bachelor businessmen moved to Creede and established new businesses. Pat Lundy, a well-known Bachelor storeowner and one of the first businessmen in the new town, was among them: "Pat Lundy, formerly of Bachelor but now located here in town, has decided to open a cash grocery store and has purchased the property on Fourth Street and is having quite a large addition built onto it by Messrs. Kennel and John York."[39] As mentioned, Lundy had purchased stock from people who had moved earlier, before he decided to move also.

Fig. 55. A log on the northeast corner of the cabin just north of the Creede Historical Society Museum. It was disassembled in Bachelor and reassembled in Creede, probably in the early 1900s. The marking shows that this was the second log on the east side. Photograph by author, 2016.

David McBride traveled to Denver and vicinity in December 1909. It was noted that Mr. McBride was one of several Bachelor residents looking for ranch homes in or near Denver.[40] In the same month, it was reported that Andy Uran had moved to Creede and was working at the Commodore Mine.[41] Like the Nelson Tunnel, the Commodore Mine had constructed a tunnel to access its mine properties from below.

While most people were moving from Bachelor to Creede, a few were moving the opposite direction: "Miss Mabel Bruns moved to Bachelor last Sunday and took charge of the boarding house there formerly conducted by Mrs. Farrell."[42]

For many old timers, the move was very difficult: "Scotty" Berry and family have left their beloved bachelor [sic] and are now living in Creede. Still, every once in awhile, "Scotty" may be seen, toiling up the hill to take a look at the old town."[43] Others left for different parts of Colorado, leaving behind friends and memories: "Miss Verna Hill and brother left the scenes of their early childhood at Bachelor to make their new home at Fort Collins. Verna will be missed by her associates of old. We hope they will enjoy their new home."[44]

Farewell parties were common in Bachelor and some were elaborate, such as the one for the Adams family in 1910.

On Friday of last week, the people of Bachelor, all friends of Mr. & Mrs. William Adams, tendered them a farewell surprise party, which certainly was a complete surprise to the household.

Mrs. Adams was conversing with a neighbor in her home during the evening. Answering a rap at the door, her astonishment was great when a crowd of her friends, male and female, filed in, all laden with good things to eat and drink. Everyone was soon enjoying themselves and fun reigned supreme.

However, another surprise was in store. A messenger arrived saying that the presence of Mr. and Mrs. Adams and those who cared to dance was requested at the Town Hall. Upon arriving there they found it illuminated and furnished with excellent piano and violin music and a few hours were pleasantly spent in dancing. When they returned to the Adam's [sic] home, the tables were found prepared with a supper that did credit to the ladies of Bachelor. The balance of the night, until the wee sma' hours were spent in various forms of entertainment, and the party broke up with expressions for good will for Mr. and Mrs. Adams from all.

Mr. and Mrs. Adams left for Denver on Thursday of this week, expecting to make that city their future home. During their six year's residence in Bachelor they have won a host of friends who wish them well.[45]

What was happening to Bachelor in the 1910s was reflected in the number of unpaid taxes on properties in the town. At a meeting of the Mineral County Board of Commissioners on September 8, 1913, the issue was highlighted by action on one Bachelor lot:

On motion duly seconded and carried that the board accept the $10.00 as payment in full for all taxes, interest, and penalties on lot 20, block 3, and improvements in Town of Bachelor, on account of depreciation of value in town being abandoned, and the same be referred to the Colorado Tax Commission for approval or rejection.[46]

Fig. 56. A portion of Bachelor in 1918 showing many apparently abandoned buildings. Several piles of lumber can be seen strewn about the site, probably the result of demolition or removal of houses. Roy Packard Collection, Creede Historical Society Archives.

Forty-five properties were included in the Delinquent Tax List for 1919.[47] In many cases, the owners probably just packed up their belongings and left because they couldn't sell their house or business (see figs. 56 and 57).

Ads for rental homes in Bachelor appeared regularly: "Four room house, furnished; nearly 300 ft. 5 ft. high, close board fencing: stable, etc. Bachelor Camp. Mrs. J. L. Wise, Address, Creede. Call at Bachelor."[48]

The last family to leave Bachelor was apparently the Allen family in 1915 (see fig. 69): "A. B. Allen and family who for sometime past have been the only family living in Bachelor, have moved to South Creede to occupy the Spangler residence."[49] This is the same Mr. Allen who killed his partner, Andy Wellington, in self-defense in 1905 and was acquitted of murder. After their parents died, the daughters, Mabel and Olive, lived for several years in Creede and supported themselves with a milk cow and a few sheep they obtained from herds moving to and from summer pastures.[50]

The last person to live in Bachelor was reported to be Annie Marshall. She was the wife of Garrett E. Marshall, a prospector. They had a son, Garrett (Gary) Marshall, who was born in 1912.[51] Gary tried to get his mother to leave Bachelor without success, so one day in 1945 or 1946 he borrowed a pickup truck to bring Mrs. Marshall and her belongings down to Creede against her will.[52]

Fig. 57. Bachelor, taken in 1922, showing the decaying town. Creede Historical Society Archives.

Teller (Bachelor) Post Office, 1894–1912

Following John Gould, the original postmaster, A. B. Gades took over on June 1, 1894. He was succeeded by S. E. Van Noorden on April 1, 1896, and Mary J. Cassedy was postmistress from August 1896 to March 1898. For some unexplained reason, there are no post office records from March 1898 until October 1903, when records reappeared. Will Wood became postmaster on October 22, 1903, and served until October 10, 1906, when Edith E. Lundy took over (fig. 58). She was postmistress through 1908 and was succeeded by Frank J. Moore from 1909 until the post office closed in 1912.[53] Biographical information on Will Wood, Edith E. Lundy, and Frank J. Moore appears in Chapter 9.

Fig. 58. Teller Post Office, 1908 or 1909. Edith E. Lundy, postmistress, is standing to the right next to an unknown girl with a dog on her back. Gardanier Collection, Creede Historical Society Archives.

Word that the post office might close became public in January 1912: "We hear rumors of the Bachelor mail route being shut off on account of lack of business. Hope it is not true."[54] The post office was officially closed on March 15, 1912.[55] The loss of the post office was the final blow for Bachelor because it represented the official judgment that there were not enough people or businesses to continue operating as a town.

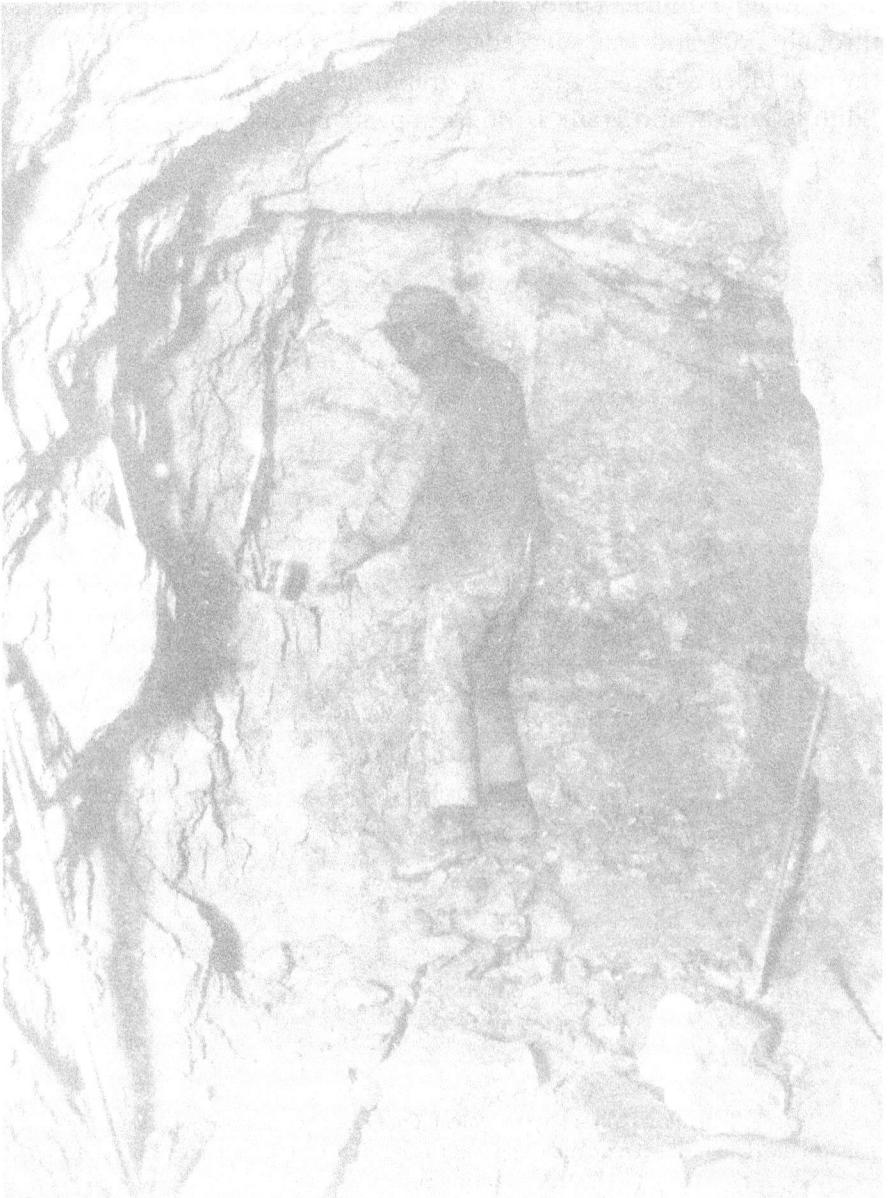

Chapter 8
Memories of Bachelor

The nominal lifetime of Bachelor, from its founding in 1892 until the post office was closed in 1912, was only twenty years, even though some people stayed there well after 1912. Despite its fleeting existence, Bachelor has a lasting memory in the hearts of people who still live in the area as well as tourists who visit the town site each summer. Its magical quality has persisted, despite the fact that it is now just an open meadow. Many people have shared their memories in publications or stories.

"Poker Alice" (Alice Tubbs) was one of the famous characters during Creede's boom days. She was a professional gambler who occasionally traveled up to the establishments in Bachelor. In a 1927 interview with *Saturday Evening Post*, over thirty years after she had left the area, she recalled her time in Bachelor:

Creede itself wasn't enough as a camp; there grew into being, high on the mountains above it, a rival city called Bachelor, where hammers clanged by night and day and life ran ceaselessly at a most turbulent pitch. Gambling halls were there, too, and a paying mine. I dealt faro and played poker in Bachelor when the predictions were common that this town would outrival Creede and for that matter form the great metropolis of Southern Colorado. There would be smelters and a railroad running over the top of the mountains, great buildings, and wide streets.

I met a friend not long ago since who had been to both Creede and Bachelor. There was no main street any more in the latter town; only a collection of tumble-down buildings with the roofs fallen in, the walls awry and the wooden sidewalks rotted back to the earth. The gambling halls where I worked, and a play of $25,000 or $30,000 a night was not at all unusual, are merely piles of rotten boards now.

In all the town during my friend's visit, only one spiral of smoke came from a chimney and that was not due to a permanent resident. A sheep herd was passing through on the way to higher country and a summer grazing. The herd crew had selected one house of the town which seemed habitable and decided to pass the night there.[1]

The abandoned town of Bachelor became a favorite site for people from Creede to visit, whether for a picnic or just to wander and reminisce. While there, it was natural to look around and perhaps find something of significance. Such was the case when Norah Dooley Korn, her mother, and two brothers hiked from Creede to Bachelor for a picnic. Mrs. Korn believes it was in 1942 and that they hiked up the steep Windy Gulch trail to get there. They picnicked near the old jail, and she remembers that it was one of the few buildings still standing. It was not very large and she recalls it was made of two-by-fours. The building was probably a place where prisoners were held for short periods of time because there were no windows, only the door. Her mother, Alma Lucille Wintz Dooley, found a key partially buried in the ground in front of the door. It was not a normal key but was hinged (fig. 59). They tried to open the door with the key but couldn't turn it and stopped trying for fear that the key might break. Norah thinks she remembers the lock being heart-shaped.[2] Mrs. Delma Dooley and her son Bill Dooley brought the key to the Creede Historical Society Museum in September 2015 for the author to examine and photograph.[3]

Many people made regular visits to Bachelor with shovels in hand, looking for the elusive bottle, old coins, or a truly rare find such as a saloon token. Others came looking for firewood or antique siding that could be used in a remodeling project in Creede or the San Luis Valley.

The saloon token in Figure 60 from the Brotzman & Long Saloon, may be one such find. It might have been found at the town site or passed down from ancestors who once lived in Bachelor. Given its condition, it could have been in the ground at one time. C. E. Brotzman and Lee Long purchased a six-month license for the saloon on March 13, 1900, for $350.00.[4]

Fig. 59. Hinged key found ca. 1942 by Alma Lucille Wintz Dooley and her family
in front of the Bachelor town jail. Courtesy, Delma Dooley and Bill Dooley
through Norah Dooley Korn.

Fig. 60. Front and back of a 12½-cent token from the Brotzman and Long Saloon
in Bachelor. Tokens from Bachelor businesses are extremely rare and highly prized
by collectors and artifact hunters. Harbert Archives

Fig. 61. Bachelor artifacts on display at the Creede Historical Society Museum. They include a mailbox door, a horseshoe, and spikes, which were found and donated to the museum by Sam Arnold. Photograph by author, 2015.

The Creede Historical Society Museum has several items that were gathered at the Bachelor town site. One set of artifacts, nicely mounted on a board that has "Bachelor, Colo." carved into it, is shown in Figure 61. The museum also has other Bachelor artifacts on display.

Jack Foster of the *Rocky Mountain News* recounted a visit to Bachelor in his "Jeep Diary" column in October 1952. He arranged to meet Fred Ryden, a Creede businessman who had grown up in Bachelor:

And it was a wonderful treat to have Fred Ryden with us. He had come from Leadville to the boisterous town of Bachelor as a kid of 6 with his parents in 1893. This was only a year after it began to boom.

So Fred had watched Bachelor blossom. And he had seen it fade. And as he now looked over all that remained of it—a few logs here, a broken piece of furniture, the remains of a board walk, and a couple of cabins still standing—he looked back through time when Bachelor was a bustling town of a thousand or two.

It was the riches of the hills—the Last Chance, Bachelor, Amethyst, Commodore—that had brought the thousands there to build the stores, drink in the saloons, pray in the churches, learn in the school.

Fred Ryden went to grammar school in Bachelor himself, from 1893 to 1904 when his family moved down to Creede. And it was a sentimental picture to watch Fred try to find the exact spot where the old school house had been. For there was nothing there now at all.

Fig. 62. Grave at Bachelor town site with aspen trees growing up through it, ca. 1960. This may be the grave mentioned by Jack Foster when he toured Bachelor in 1952. Courtesy, Denver Public Library.

Foster kept narrating as he and Ryden continued on their nostalgic journey (fig. 62):

And Fred showed us, too, where his own house had stood. And he remembers that the Allen house stood not far away. One evening he saw a man named Wellington walk to the Allen house with a gun and Allen shot him dead because he was annoying his daughters.

"Wellington was buried in the Bachelor cemetery," said Fred. "And it's somewhere there among the aspen."

My companion and I drifted down among the aspen. But we could not find the cemetery until Fred had shown us exactly where it was. For it was grown over now and most of the wooden head-pieces are warped and nameless. Where Wellington lies no longer is marked, nor are scores of other rock mounds great and small. They lie there as the nameless end of life.

Only one grave is marked. It is surrounded by a fallen wooden fence, and faintly on the wooden head-piece is printed this memory: "John Erskin. Died March 23rd, 1893."

And through his body a tall aspen now is growing, and a rat's nest is all but hidden nearby.

"This was a great town when I was a child," Fred recalled as we walked up the grassy stretch that was once a street. "The ore was rich, the stakes were high, and the men were young. And there was a good life, too, and we had lots of fun."

"But after the turn of the century, the ore began to pay out and people began to go away. Now, as you can see, there is no one left."[5]

Muriel Sibell Wolle, a University of Colorado professor, toured the backroads of Colorado visiting former mining and ghost towns throughout the state. She cataloged her travels in *Stampede to Timberline*, a popular reference book for people traveling through or researching ghost sites in Colorado. She was an accomplished artist and illustrated the book with sketches of many of the towns she visited. Surprisingly, she was unable to find Bachelor on her first trip, around 1960. She had been told by the Creede postmaster that wood sidewalks were about all that was left (fig. 63):

Fig. 63. Remnants of the wooden sidewalks or a fallen fence at the Bachelor town site, 1960 to 1970. Courtesy, Denver Public Library.

The trip up Bachelor hill, although only three miles, was stiff. Partway up, the road passed the cemetery and nearer the top it ran beside a rocky gulch, down which tiny foot trails ran to prospect holes and mine tunnels. The road wound on past the end of a meadow, and still there were no sidewalks. I drove another mile over a mere sliver of trail until it ended among some mine buildings. This wasn't Bachelor, and since there was no one around to question, I returned to Creede.[6]

She came back the following year and had more success:

Victoria and I scanned every gully, meadow and patch of timber for the wooden sidewalks, but found nothing. Parking the car where Bachelor should be, I climbed up one side of a hill while Victoria climbed up the other. On the edge of a grove of aspens was a government stake marked "Bachelor Mountain." This was encouraging, so I went on. Hearing a shout in the distance I hurried to an opening through the trees and saw, in the next clearing, two long rows of splintered planks—the sidewalks of the town. They bordered a gully which was once the main street. A few cabins were standing, though most were toppled over or completely flattened by the weight of successive snows. A few pieces of furniture, including a round-topped trunk, were strewn amidst the general debris. A table, its four legs in the air, lay half covered with the mud which had washed over it in the spring rains. . . . An occasional stone terrace or a fragment of picket fence showed where homes once stood.[7]

Another book on ghost towns, by Caroline Bancroft, also describes Bachelor's condition during the same period Professor Wolle visited the town:

In 1960 there were only three cabins left standing on what was formerly Bachelor's residential street and a few remnants of the boardwalk on its main street. Among the trees on the east side of the meadow, where Bachelor once lay, was a narrow picket-fenced grave, shaded by trees. A local story says that three bodies are buried there, one on top of the other, because of the difficulty of digging in frozen ground the day after the tragedy that claimed all three.[8]

Sheila Goodman wrote an article in the *Mineral County Miner* in 2006 that summed up Bachelor's decline (fig. 64):

Bachelor was short-lived as far as towns go with many residents moving their small houses into the Creede area. The houses left behind were used as building material or firewood by the few remaining town folk. Eventually, they also left, leaving the buildings to rot or be torn down and then hauled away by people in the area and by artists and crafters as well as collectors.[9]

Fig. 64. Old houses in Bachelor, ca. 1960. Courtesy, Denver Public Library.

In preparation for this book, the author walked the Bachelor town site on several occasions in 2015. On one trip, former Mineral County sheriff Phil Leggitt gave the author a tour of the site. Other than the modern home on private land at the bottom of the old town, very little was visible from the road. There are remnants of one or two old houses among the aspen groves on the west side of the town; they can't be seen from the road or the parking area (fig. 65). The main street is now a gully with an occasional piece of wood that may have been from the sidewalks.

Part of an old wood stove is out in the open about two-thirds of the way down the hill (fig. 66). Nearby, there is also an old bed spring rusting in the meadow. Many rusty tin cans and shards of glass and ceramics can also be seen in various places; depressions in the dirt can be seen throughout the town site. George Carpenter, a former Creede resident, told the author that many of the depressions are former outhouses because they were favorite digging sites.[10]

Fig. 65. Remains of cabin in aspens on west side of Bachelor town site, summer 2015. Photograph by author.

Fig. 66. Remnants of what appears to be a wood stove at the Bachelor town site. It is below the cabin ruins shown in figure 67, near the bottom of the town. Photograph by author, 2015.

From the road, the most visible building is the ruin of a former cabin. It is located in the lower section of town, a few hundred feet above the modern home. The most prominent feature is a vertical wooden pole or beam sticking up above the building (fig. 67).

Fig. 67. Remains of a cabin in the lower portion of the former town. Photograph by author, 2015.

In addition to the many photographs of Bachelor in decay, local artists have captured views of the site on paper or canvas. One was Toni Davlin, an artist and also co-founder of the Creede Museum.[11] She did several sketches of local mines and Bachelor. One is shown in figure 68.

Fig. 68. Sketch of one of the decaying cabins in Bachelor (ca. 1940s) by Toni Davlin. Courtesy, Doug Davlin.

Bachelor has many wonderful memories. It is a special historic site that has been enjoyed by many Creede residents and visitors to the Creede area. It will certainly be visited by many more people in the future. Please enjoy the historic town site, but treat it with the respect it deserves so others may enjoy it in the future.

Chapter 9
People of Bachelor

From the original founders to some of the last people who lived in Bachelor, a long list of Bachelor residents deserve special mention. Some were respected members of the town, while others are included because of their notoriety. This is a compilation of brief descriptions for some of the notable residents of the town. Many were discussed in greater detail earlier in the book.

The Allen Family. Arthur Allen and his wife Amanda were from England. They had three children—two daughters and a son—born in Colorado (fig. 69). The Allen family is listed in the 1900 and 1910 U. S. Censuses of Bachelor. Mr. Allen was very protective of his daughters and killed Andy Wellington, probably after the latter made unwanted advances toward them. The Allens were reported to be the last family to leave Bachelor, moving to South Creede in 1915. After the parents died, daughters Mabel and Olive raised a milk cow and delivered milk to residents of Creede.

Scotty Berry was a shift boss at the Last Chance Mine. He greeted miners each morning with "AH and 'tis Scotty Berry that has the prestige this morning!"[1] He had a reputation for being the life of the party, as noted in a 1905 issue of the Creede Candle: "Scotty Berry returned from Pueblo Thursday and now there will be more life in Bachelor."[2]

Dr. John A. Biles was the Bachelor town physician for many years. He was a southerner who rode a black horse and wore a matching black hat. He reportedly carried a black bag that contained his tools and a bottle of whiskey. As with most physicians of that era, his office and operating room moved from one patient's house to another.[3] He was listed as a Bachelor resident in the 1900 U. S. Census, but not in the 1910 Census.

Sam D. Coffin was an early prospector in the area, having arrived in December 1890 (fig. 15). Along with John MacKenzie and Nicholas Creede, he is generally regarded as one of the original prospectors in the Creede Mining District.[4] He discovered many claims on Bachelor Mountain and built one of the first homes in Bachelor. Some of his claims

Fig. 69. Arthur and Amanda Allen with their children Mabel, Olive, and Archie.
Creede Historical Society Archives.

were on the Bachelor town site. He was one of the original thirty-three petitioners on the April 13, 1892, request for incorporation of Bachelor and was among the 122 men who participated in the May 25, 1892, vote for incorporation.

A. S. Crawford was one of the original petitioners for incorporation of Bachelor. He moved to Bachelor from Leadville and was selected to be on the town's first Board of Trustees. He owned and operated a sawmill near the town.

Jack Dempsey was the most famous person who lived in Bachelor. He moved with his family from Manassa, Colorado, to Bachelor in 1900

Fig. 70. Jack Dempsey home (marked with X) in Bachelor. Courtesy, Creede Historical Society Archives.

Fig. 71. Andy Dooley. Creede Historical Society Archives.

(fig. 70). They lived near the William T. Jackson home before they moved to the Montrose area, where Dempsey began his boxing career.[6] He was the world heavyweight box champion from 1919 to 1926.

Andy Dooley is the patriarch of the many Dooleys who have lived in/or still live in Creede and Mineral County (fig. 71). He was a miner at the Last Chance Mine. Because he was so tall, he is easily recognizable in photographs of miners in the early 1900s.[7] He and his wife, Bridgette, moved to Bachelor in 1908 and then apparently moved to Creede before 1910, because they are not listed in the 1910 Bachelor census.

Asa B. (Cap) Eades owned a grocery store, one of the last businesses in Bachelor. His is one of only five businesses listed in the *1911 Colorado Business Directory*. According to Harold Wheeler:

He had two daughters, Alda and Hulda. Alda ran off with a gambler. Hulda was my own age. She used to furnish us with Bill [sic] Durham and papers from the store. Hulda was never with our gang much. Whenever she got out of sight, her mother could be heard all over the county shouting, 'HULDAAaa.' With the remarkable echo we had, each shout was magnified. You could always tell when meal time was coming up. Mrs. Eades broke up a wooden box to start the kitchen fire.[8]

Joseph Either was Ed O'Kelley's accomplice in the murder of Bob Ford. He was one of the thirty-three Bachelor founders who requested incorporations in April 1892 and also participated in the incorporation vote. He and O'Kelley conducted the Bachelor census to support the incorporation request.

Ed "the Barber" Fulst was born in Germany and ran a barbershop in Bachelor. He apparently came to Bachelor in the early 1900s because he is listed in the 1910 U. S. Census but not in the 1900 Census. Because of his heavy German accent, he was frequently the target of practical jokes. He had a dog named Kaiser and apparently had a severe temper. During a drunken spell when he was incapacitated, a group of Bachelor men shaved the dog to look like a lion and also shaved off half of Ed's "Kaiser Bill" mustache. When he awakened and discovered the mischief, he grabbed his pistol and went to Main Street, firing in all directions.[9] Despite his temper, he was elected mayor of Bachelor in 1906. While serving as mayor, he was arrested and later acquitted for taking a shot at two deputy sheriffs who were trying to arrest a drunken man in town.

Charles Goodman was a well-known photographer who had a studio in Bachelor. He took many outstanding photographs of Bachelor and the surrounding mines (see figs. 31 and 40). He was called an artist by the *Teller Topics*.

John Gould was a member of Bachelor's first Board of Trustees as well as the town's first postmaster. He owned and operated the Pioneer Store, which sold groceries, clothing, hardware, and mining supplies (see fig. 35).

Gustav Hoffman was the first mayor of Bachelor. What is surprising is that he was not one of the original thirty-three petitioners for incorporation on April 13, 1892, and he did not vote in the incorporation election on May 25, 1892. Regardless, he was respected enough to become the inaugural mayor. He owned a hardware and notions store that was listed in the *1893 Colorado Business Directory* (see fig. 37). He is not listed in the 1900 Bachelor census, so he apparently left the town in the 1890s.

William T. Jackson, Sr. moved with his family to Bachelor in the 1890s and worked at the Last Chance and Amethyst Mines. He is the patriarch of the family that included many Creede miners, including his son, William T. Jackson, Jr., and his grandsons, John R. Jackson and William C. Jackson. He is listed in the 1900 and 1910 Bachelor censuses. He later moved to Stringtown (between North Creede and Creede) and died in the early 1950s.[10]

J. W. Jenkins was one of the thirty-three petitioners for Bachelor's incorporation and was elected to the town's first Board of Trustees. He is listed as owner of the Gem Saloon in the *1893 Colorado Business Directory*.

Edith Eleanor Oberg Lundy came to Bachelor with her mother, Annie Johnson Oberg, who married Pat Lundy in 1895. Edith was postmistress of Bachelor from 1906 through 1908, and would frequently ride on horseback to Creede to meet the train and pick up the mail (fig. 72). She married Sutter A. Gardanier in 1909 and moved to Creede.[11]

Pat Lundy ran a grocery store in Bachelor. He is also listed under the Lundy and Sherry Saloon in the *1893 Colorado Business Directory*. He married Annie Johnson Oberg in April 1895. According to Harold Wheeler, Lundy was quite a storyteller. Edwin Lewis Bennett confirmed this and added that he ended each tall tale with "and the pretty part of it is it's the truth."[12] He was the step-father of Edith Eleanor Oberg Lundy. Lundy was a longtime resident who was listed as a Bachelor resident in

Fig. 72. Edith Eleanor Oberg Lundy (right) in 1908 meeting the train in Creede to pick up the mail for Bachelor. Courtesy, Larry Gardanier.

the 1900 and 1910 U. S. Censuses. He was listed as owner of a general merchandise store in the *1911 Colorado Business Directory*.

George C. Martindale was one of the petitioners for the incorporation of Bachelor. Although not on the original Board of Trustees, he is listed as a City Council member in the *1893 Colorado Business Directory*. He apparently left Bachelor in the 1890s because he is not listed in the 1900 U. S. Census.

John C. MacKenzie was the first prospector to arrive in the area (fig. 13). He established his base in Sunnyside but must have traveled many times through the mountain park that later became Bachelor. He discovered the Bachelor claim in 1884, but for unknown reasons he did not develop the mine until much later. He and Sam D. Coffin arranged for the original plat of Bachelor. It is appropriate that he is regarded as the "Father of Bachelor."

Frank J. Moore was a late arrival to Bachelor. He first appears in the *1908 Colorado State Business Directory* as co-owner of a saloon. He was postmaster from 1909 until the post office closed in 1912.

I. W. Newland was one of the thirty-three petitioners for the incorporation of Bachelor and was elected to the first Board of Trustees.

Ed O'Kelley became Bachelor's most notorious resident when he killed Bob Ford in Creede on June 8, 1892. Ford gained fame when he killed Jesse James in April 1882, ten years before he came to Creede. O'Kelley was one of the founders of Bachelor, participating in the request for incorporation and in the incorporation vote. He was serving as Bachelor town marshal when he killed Ford.

C. H. Pierson was another of Bachelor's founders. He has been credited with establishing the first business in the town—a livery, feed, and transfer business that advertised two daily trips to Jimtown (fig. 33). Pierson was also elected to the first Board of Trustees.

Tom W. Vincent was another of the founders of Bachelor and was the first town clerk and treasurer. He owned a real estate business and also sold lumber and coal (see fig. 34). He was elected to the first Board of Trustees.

Harold French Wheeler grew up in Bachelor in the early 1900s and shared many of his childhood experiences in his family memoir—a valuable historical legacy. He moved with his family to Creede, where he met and married Muriel Belle LeZotte in 1921. Many of his colorful descriptions of Bachelor residents and experiences are included in this book.

A. H. Whitehead was elected to the first Board of Trustees of Bachelor despite the fact that he apparently did not participate in the incorporation petition or vote. He owned the Teller House (see fig. 36), which advertised "first class accommodations." He is also listed as a justice of the peace in the *1896 Colorado Business Directory.*

Will Wood and his wife, Mae, moved from Gunnison to Bachelor in early 1898 (fig. 73). According to a poster at the Creede post office, he was postmaster of Bachelor from October 22, 1903, to October 10, 1906. However, he is also listed as postmaster in the *1902 Colorado Business Directory.* Wood also owned a grocery and meat business. He and his family moved to Olathe, Colorado, in September 1906.[13]

Fig. 73. Will Wood, businessman and postmaster of Bachelor. Courtesy, Lewis (Bud) J. Wood, 2016.

Epilogue

This has been the most difficult and enjoyable of the four books the author has written in terms of the amount and depth of research required. Fortunately, there was a lot of help along the way. The author was pleasantly surprised to find that the historical material is available if one is willing to dig and have people to assist you.

The availability of issues of *Teller Topics* at the Stephen H. Hart Library in Denver, Colorado, was a significant discovery, even though the issues covered only a few months in 1892. Perhaps the greatest surprise was finding the original incorporation documents at the Hinsdale County Museum. The author appreciated the assistance he received from Grant Houston.

The most difficult part of the project is knowing that additional reference material is out there—the author couldn't find it despite a diligent search. Hopefully, publication of this book will stimulate additional information coming to light.

When the author moved to the Creede area in 1999, he was surprised and disappointed to find that so few buildings were still standing in Bachelor. He has toured many ghost towns in Colorado where buildings from the late 1800s are still standing. It is now apparent from his research that many Bachelor buildings were moved to Creede and others, such as the Congregational Church, were torn down for building materials. Almost certainly, others were destroyed and used as firewood to provide warmth during the cold winters. These actions were enhanced by the close proximity of Creede. Natural decay claimed buildings that were not torn down or moved. Regardless, it is a shame that more of the town was not preserved for historical purposes.

The author was surprised and pleased to get into the essence of Bachelor in a profound way. He could easily imagine walking down Main Street and stopping to talk with John MacKenzie, Sam Coffin, Gustav Hoffman, Will Wood, Edith Oberg Lundy, Andy Dooley, Ed Fulst, and all of the other wonderful residents of the town. It would have been fascinating to meet Ed O'Kelley before he murdered Bob Ford. Having lived in Mineral County for nearly twenty years, the auther understands how difficult life must have been—particularly in the winter.

It has been an interesting journey. The author is sure there will be many times in the future—at the Creede Historical Society Museum or elsewhere—when memories come streaming back and new information comes to light.

The book ends with the beautiful and haunting poem written by Jane Morton and published in the 2012 issue of the *Willow Creek Journal*. The author first heard it at the *Mining through Poems, Songs, and Stories* series organized by Johanna Gray and sponsored by the Creede Historical Society. The poem captures the full range of emotion of this wonderful, if short-lived, town. We can't stop the inevitable decay, but we can preserve some of its memories.

Bachelor (1892–1910)

Ten thousand and five hundred feet
    above sea level lies
The Bachelor town site peaceful now
    years after its demise.
It had been a natural park
    a grassy mountain lea—
flat enough to build upon
    beyond the mine debris.
The working mines were located
    the far side of the hill
where men could walk from Bachelor town
    to work in mine or mill.
More bachelors than married men
    each day strode to and fro,
And thus the name described the place,
    which seems most apropos.
The building went on night and day.
Men hammered, sawed, and jawed.
Log cabins, shanties, businesses
    sprang up above the sod.
A thousand people at its peak—
    One eighty-acre site.
Perhaps close quarters may have caused
    hot tempers to ignite.
This new town generated noise.
    It partied through the night,
carousing, drinking, gambling

until the sky turned light.
Fist fights broke out in the saloons,
    attracting quite a crowd,
which soon moved out into the street,
    and turned unruly, loud.
Occasional horses galloped by.
    Freight wagons lumbered in.
Loud gunshots went off anytime
    to sleepless folks' chagrin.
This hustling, bustling mining town
    thrived during mines' heyday.
Men spent their off-shift hours there,
    along with hard-earned pay.
The boarding houses, bordellos,
    saloons and general stores,
strived to meet the needs of those
    whose lives involved the ores.
As went the mines, so went the town.
    In eighteen-ninety-three
the Sherman Silver Act repeal
    near wrecked the industry.
It didn't touch Creede mines so much,
    at least not right away.
Within the year, though, mines cut back
    The jobless moved away.
Then in the bowels of the mines
    the water rose too high.
Since pumping wasn't feasible
    they gave tunnels a try.
But when it wasn't profitable
    to keep producing ore.
Mines had to shut production down,
    and Bachelor was no more.
Now smoke stacks rust down near the mines.
    No blasting breaks the spell.
Amid huge piles of fractured rock
    ground squirrels and marmots dwell.
The wind blows down the Bachelor streets
    and crosses vacant lots.
It dusts off old foundation stones.
    What's left of boardwalk rots.
The town itself has disappeared,
    its structures razed for wood,
or else moved down to Creede below
    from lots on which they stood.
The townsfolk who were left behind

Are sleeping on the hill.
They sleep beneath the aspen trees,
And all is peaceful, still.

Jane Morton (2012)

Appendix 1
Ordinances of the Town of Bachelor

At their regularly scheduled meeting on Tuesday, July 19, 1892, Mayor Hoffman and Trustees Pierson, Gould, Jenkins, Newland, Likens,[1] and Whitehead approved several actions. The most important was passage of the sixteen ordinances that formed the legal foundation for the governance of Bachelor. The ordinances were approved and, by unanimous vote, the town officers passed a motion to have them published in the town's new newspaper, *Teller Topics*. They were almost certainly patterned after the Creede town ordinances because Ordinance 14 was initially published with Creede inserted in the last sentence of Section 1. It was subsequently corrected to Bachelor.

The fact that the town would go to the trouble to write and publish its own ordinances in its infancy was evidence that the leaders believed in Bachelor's future. It also demonstrated that they believed the town would grow into a large city requiring the ordinances for effective governance.

The vote of the trustees was reported in the first edition of the *Teller Topics*, published on July 22, 1892: "Sixteen ordinances needed for the government of the city were passed. It was unanimously voted to have all ordinances published in the *Teller Topics*, consequently they appear in this issue and speak for themselves."[2]

The first fifteen ordinances were published in the July 22, 1892, issue of the *Teller Topics*. Because of an error, ordinance 14 was reprinted in the July 30, 1892, issue. Ordinance 15, published in the July 22, 1892, issue, concerned accounts. A different ordinance 15, concerning dogs, was published in the August 6 issue. While the total is sixteen, it is not certain whether the second ordinance 15 was intended to become ordinance 16.

The approval of town ordinances was another important step forward in bringing order to the new town. The ordinances listed the responsibilities of each member of the local government, addressed public safety issues such as law enforcement and fire protection, and provided enforcement mechanisms.

The ordinances are reproduced in their entirety here to allow the reader to see the template by which the town was governed. Each of the ordinances was signed by the new mayor, Gustav Hoffman, and attested by the town recorder, Tom W. Vincent. The town seal, whose design is specified in town ordinance 6, was affixed to each ordinance (see fig 74).

Ordinances of the Town of Bachelor.
SERIES OF 1892, NO. 1.

An ordinance establishing rules and order of business.

Be it enacted by the trustees of the town of Bachelor:

Section 1. That the rules and order of business shall be as follows:

Rule 1. The rules of procedure and order of business shall be strictly adhered to by the board of trustees, unless temporarily suspended by a two-thirds vote of the members present.

Rule 2. The board shall meet regularly on the first Tuesday and third Tuesday of each month at 7 p.m., or at any other day and hour that the trustees in their discretion may deem proper.

Rule 3. The mayor and any two members may call at any time special meetings of the board.

Rule 4. The mayor, or in his absence, one of the trustees who may be elected mayor pro tempore, shall preside at all meetings of the board of trustees.

Rule 5. The board shall, at its first meeting, or as soon thereafter as practicable, choose one of the trustees as mayor pro tempore, who, in the absence of the mayor from any meeting of the board, or during the mayor's absence from the town or his inability to act, shall act as mayor.

Rule. 6. Four members shall constitute a quorum to transact business, but a minority may adjourn from time to time and compel the attendance of members.

Rule 7. At the hour appointed for the meeting, the members shall be called to order by the presiding officer, who shall order the roll call, note the absentees and announce whether a quorum be present, and if there be a quorum, the board shall proceed to business in the following manner to-wit:

1. Reading of minutes.
2. Presentation of petition.
3. Reports of officers.
4. Reports of standing committees.
5. Reports of select committees.
6. Reading of bills.
7. Unfinished business.
8. Introduction and consideration of ordinances.
9. Motions, resolutions, inquiries and appointments.

Rule 8. All questions relating to priority of business shall be decided without debate.

Rule 9. The presiding officer shall preserve order and decorum, and shall decide all questions of order, subject, however, to the right of appeal to the board.

Rule 10. Previous to his speaking, every member shall address the presiding officer, and shall not proceed with his remarks until recognized by the chairman. He shall not speak more than twice on each subject without permission of the board. No member shall leave the room while the board is in session without permission of the presiding officer. Any member called to order shall immediately suspend his remarks and resume his seat unless permitted to explain.

Rule 11. A motion to consider a vote shall only be made at the same meeting, or the next succeeding meeting, provided no action shall have been taken on the same.

Rule 12. All motions and resolutions shall be reduced to writing if required by any member of the board, and when seconded and stated by the chair, or read by the clerk, shall be open for consideration. No motion or resolution can be withdrawn after it has been decided.

Rule 13. A motion to adjourn shall be in order, shall have precedence over all other motions and shall be decided without debate.

Rule 14. Every member present shall be required to vote on all questions, except when excused by the board. The ayes and noes shall be called and the names of the persons voting shall be recorded in the minutes. All appointments to office shall be by ballot, and a majority of all members elected and qualified shall be necessary to a choice. All committees shall be appointed by the presiding officer unless otherwise ordered by a two-thirds vote of the board, and in that case they shall be appointed by ballot.

Rule 15. All reports of committees shall be in writing and be address to the board of trustees of the town of Bachelor.

Rule 16. The standing committees shall be appointed annually, and the person first named on the committee shall be the chairman thereof. Each committee shall consist of three members. The following shall be the standing committees and others, to-wit:

1. Committee on finance, printing and licenses
2. Committee on streets, alleys and bridges
3. Committee on fire, water, lights and public buildings.

Rule 17. When the board is in executive session, the room shall be cleared of all persons except town officers. All proceedings in executive session shall be kept secret.

Rule 18. The board shall make and enforce the rule of procedure.

Rule 19. The board shall keep a journal of its proceedings.

Rule 20. If the question in debate contains several distinct propositions any member may have the same divided.

Rule 21. The previous question shall be as follows: "Shall the main question now be put?"

Rule 22. While a member is speaking no member shall maintain private discourse.

Rule 23. While the presiding officer is putting the question no member shall walk across or out of the council room.

Rule 24. When a blank is to be filled and sums and times proposed, the question shall first be put upon the the [*sic*] greatest sum and lougest [*sic*] time.

Passed by the board of trustees and approved by me this 19th day of July, 1892.
GUSTAV HOFFMAN, Mayor.
Attest: T. W. VINCENT, Recorder.[3]
[SEAL.]

SERIES OF 1892, NO. 2.

An ordinance regulating the appointment of town treasurer and other subordinate officers, and prescribing duties not specially defined by statute.

ARTICLE 1

Be it ordained by the board of trustees of the town of Bachelor:

Section 1. That there shall be appointed by the board of trustees on 10 days after the town election, or as soon thereafter as practicable, a town treasurer, a town recorder, a town attorney and a town marshal; and said officers shall hold their respective offices for one year and until their successors are appointed and qualified, unless sooner removed according to law and hereinafter provided.

Sec. 2. Before entering upon the duties of their respective offices, each officer mentioned in this article shall take and subscribe an oath or affirmation to support the constitution of the United States and the constitution of this state and the laws of this state made in pursuance thereof, which oath shall be filed with the town recorder.

Sec. 3. Any officer named in section one of this ordinance may be removed for cause by the concurrent votes of four members of the board of trustees. No such removal shall be made however, without a charge in writing and an opportunity of hearing being given to said officer.

Sec. 4. Any officer elected or appointed who shall fail or neglect any duty imposed on him by the ordinances of the town of Bachelor, unless therein otherwise provided, shall for each and every such offense be fined or pay a penalty of not less than $10 nor exceeding $25.

ARTICLE II.
TOWN RECORDER

Section 1. The town recorder shall be the keeper of the town seal and shall affix it to all papers and instruments which by law or ordinance are required to be attested by the town seal. He shall have the custody of and shall safely keep all public documents, resolutions, records, ordinances and orders for the board of trustees and such other papers and documents as may be delivered into his custody.

Sec. 2. It shall be the duty of the town recorder to attend all meetings of the board of trustees, keep the minutes of all their proceedings, and record the same in the books to be provided by the board of trustees and to be kept in his office. He shall keep a correct account between the town and the town treasurer, by charging him with all the sums received by him as exhibited to the town recorder in his duplicate receipts and crediting him with all moneys paid out by him by order of the board of trustees; and he shall allow the town treasurer such other credits as he may be entitled to by law and ordinances of the town. He shall countersign all warrants drawn on the treasury and deliver the same when called for, taking a receipt therefor, and generally he shall do and perform such other duties as may from time to time be enjoined upon him by ordinance or resolution by the board of trustees. It shall be the duty of the town recorder at the close of the fiscal year, to make out and lay before the board of trustees a full and explicit statement of the receipts and expenditures of all the fiscal affairs of the town during such year, and cause the same to be published in such manner as the board of trustees may direct.

Sec. 3. It shall be the duty of the town recorder to keep an accurate account, charging each item to its proper appropriation, as set forth in the annual appropriation ordinance, and to keep such accounts in such a manner that he may be able at any meeting of the board of trustees to report the credit upon his books to any appropriation.

Sec. 4. The clerk and recorder shall receive for his services such compensation as the town board and recorder shall arrange upon, and also such fees as may be provided by ordinance or resolution.

ARTICLE III.
TOWN ATTORNEY

Section 1. The town attorney shall draft all ordinances, leases, conveyances and all instruments of writing which may be required of him by ordinance, motion or orders of the board of trustees.

It shall be his duty to act as legal adviser to the town in all matters pertaining to contracts with or by the town, or in any question of law arising under any ordinance or otherwise.

Sec. 2. The town attorney shall have the right to be heard on all questions or motions before the board of trustees, amending, repealing or in any manner affecting any ordinance in force or to be enacted by the board of trustees, and shall receive for his services such compensation as he and the board of trustees may agree upon and fix by contract or resolution.

ARTICLE IV.
TOWN MARSHAL.

Section 1. The town marshal shall have the same power that constables have by law, co-extensive within the county in cases of violation of town ordinances and for offenses committed within the town.

Sec. 2. The marshal shall perform all the duties required by the ordinances of the town, or any of them, and shall execute all orders given by the mayor, town attorney, town recorder or chairman of standing committees in their official capacity. He shall devote all his time to the performance of his duties. He shall attend all meetings of the board and shall serve all papers ordered to be served by the officers of the town in their official capacity. He shall serve notices of special meetings upon each member of the board and upon the town attorney personally or by the leaving the same at his residence or place of business with some responsible person, at least twelve hours previous to said meeting.

Sec. 3. The town marshal shall be master of the pound of the town of Bachelor.

Sec. 4. The town marshal shall, upon appointment, and before entering upon the discharge of his duties, execute to the people of the state of Colorado a bond, with two good and sufficient sureties, to be approved by the board of trustees of the town of
Bachelor, in the penal sum of five hundred dollars conditioned for the faithful discharge of the duties of the office.

Sec. 5. It is hereby made a misdemeanor for the town marshal to enter upon the discharge of the duties of his office until said bond has been executed, approved and filed with the clerk and recorder of the town, and upon conviction shall be fined in any sum not more than twenty-five dollars.

Sec. 6. The town marshal shall receive such monthly compensation as shall hereafter be agreed upon by ordinance or resolution.
He shall receive in addition for the arrest of each person convicted of a violation of the ordinance of the town of Bachelor such fees as constables now by law receive in counties of the fifth class.

ARTICLE V.
TREASURER

Section 1. The treasurer shall give a bond to the town of Bachelor, with good and sufficient sureties, to be approved by the board of trustees, in the sum of five thousand dollars, and conditioned for the faithful performance of his duties as treasurer, and that when he shall vacate such office, he will turn over and deliver to his legal successor in office all moneys, books, papers, property or things belonging to the town and remaining in his charge as treasurer.

Sec. 2. The treasurer shall receive all moneys belonging to the corporation, and shall use the same methods in keeping his books and accounts as are used by competent bookkeepers; and such books and accounts shall always be subject to inspection by any member of the board of trustees at reasonable hours.

Sec. 3. He shall keep a separate account for each fund or appropriation, and the debts and credits belonging thereto. He shall give every person paying money into the treasury a receipt therefor, specifying the date of payment and upon what account paid; and he shall also file statements of such receipts with the clerk at the date of his monthly reports.

Sec. 4. He shall, at the end of each and every month, and oftener if required, render an account to the board of trustees or to such officer as may be designated by ordinance, showing the state of the treasury at the date of such account, and the balance of money in the treasury. He shall also accompany such accounts with the statement of all moneys received into the treasury, and on what account, during the preceding month, together with all warrants redeemed and paid by him; such warrants, with any and all vouchers received by him, shall be delivered to the clerk and filed with his account in the clerk's office upon every day of such statement.

He shall return all warrants paid and by him stamped or marked "paid." He shall keep a register of all warrants redeemed and paid which shall describe such warrants and show the date, amount, number and fund from which paid, and the name of the person to whom paid.

Sec. 5. He shall keep all moneys belonging to the corporation in his hands separate and distinct from his own moneys, and he is hereby prohibited from using, either directly or indirectly, the corporation money or warrants in his custody and keeping, for his own use and benefit or that of any other person or persons whomsoever, and any violation of this provision shall subject him to immediate removal from office by the board of trustees, who are hereby authorized to declare said office vacant, and in which case his successor shall be appointed, who shall hold his office the remainder of the unexpired term.

Sec. 6. The treasurer shall annually, between the 18th and 28th days of July, make out and file with the clerk a full and detailed account of all receipts and expenditures, and of all his transactions as treasurer during the preceding fiscal year, and shall show in such account the state of the treasury at the close of the fiscal year, which account the clerk shall cause to be immediately published by having the same printed in some newspaper published in the town of Bachelor.

Passed by the board of trustees and approved by me this 19th day of July, 1892.

GUSTAV HOFFMAN, Mayor.

Attest: T. W. VINCENT, Recorder.[4]
[SEAL.]

SERIES OF 1892, NO. 3.

An ordinance concerning licenses.

Be it ordained by the board of trustees of the town of Bachelor:

Section 1. That licenses may be issued in said town, subject to the ordinance which may be in force at the time of the issuing thereof, or which may subsequently be passed, and if any person so licensed shall violate any of the provisions of his license or bond, he shall be liable to be proceeded against in the manner hereinafter provided, and his license shall be revoked, in the discretion of the board of trustees.

Sec. 2. No license shall be granted for a less period than six months, nor more than one year, nor shall any license be transferrable without permission from the board of trustees.

Sec. 3. All licenses shall be issued and signed by the mayor and recorder under the town seal, and dated as of the day of application therefor, upon payment of the sum assessed and the fees of the town recorder for issuing the same.

Sec. 4. The town recorder shall keep a license register, in which shall be entered the name of each and every person licensed, the date of the license, the purpose for which granted, the amount paid therefor, and the date at which the same expires.

Sec. 5. Any person or persons who shall keep for public use within the limits of this town any ball alley, billiard, bagatelle or pigeon-hole table without first having obtained a license shall forfeit and pay to the said town the sum of five dollars for each and every offense.

Sec. 6. The license issued shall be signed by the mayor and recorder, who shall issue license to persons applying therefor for the purposes specified in the foregoing section upon the payment by the applicant of the sum of five dollars for each quarter of a year, or twenty dollars a year for each and every ball alley, billiard, bagatelle or pigeon-hole table, and at the same rate for each additional ball alley, billiard, bagatelle or pigeon-hole table.

Sec. 7. Any person or persons who shall have taken out license for any of the purposes named in the foregoing section who shall allow any person under the age of sixteen years to frequent the room where such ball alley, billiard, bagatelle or pigeon-hole table is situated, shall upon the complaint of the parent or guardian of such minor, forfeit and pay to said town for each offense the sum of twenty dollars.

Sec. 8. The mayor and recorder are hereby authorized to grant licenses for the sale of spirituous, vinous, malt, fermented and ardent liquors to any person who shall apply therefor in writing, upon the payment of five hundred dollars for each year from and after the passage of this ordinance; also, to grant licenses for the sale of malt or fermented liquors only upon the payment by the applicant therefor of the sum of two hundred and fifty dollars for each half year.

Provided, That the person applying for any such license shall, before the issuance of the same, execute to the town of Bachelor, a bond with two sufficient sureties, to be approved by the mayor and recorder, in the penal sum of two thousand dollars, conditioned that the parties so licensed shall keep an orderly house and shall faithfully keep and observe all ordinances in force in said town during the period of such license regulating and governing the keeping and keepers of saloons and dramshops.

Provided further, That no license shall be issued for a less period than six months, and that such license shall not be assignable and shall not authorize the person or persons therein named to sell, barter or give away any such liquors at more than one place or house in said town.

The license so granted shall authorize the person or persons therein named to sell, give away and deliver any of the liquors specified in this section as applicable to each of the said licenses above authorized in quantities of less than one gallon at the place designated in the application.

PEDDLERS AND HAWKERS

Sec. 9. Any person who shall pursue the occupation of peddler or hawker within the town limits without first having obtained a license therefor, or who, having obtained such license, shall be guilty of any fraud or deceit in the conduct of his business, or shall in any way violate any of the provisions of his license, shall for each offense forfeit and pay to said town a sum of not less than five nor more than fifty dollars.

AUTHORITY–LIMITATIONS.

Sec. 10. The mayor and recorder is [sic] hereby authorized to issue to any person applying therefor a license for the purpose named in the foregoing section, upon said applicant paying the sum of fifty dollars for each quarter of a year. Such license shall set forth as clearly as possible the kind and aggregate value of the goods, wares and merchandise to be sold by virtue thereof, and the mode of conveyance of the same, whether by cart, wagon, truck or otherwise.

PERSONS EXEMPT.

Sec. 11. Persons coming into this town with teams or otherwise for the purposes of selling vegetables or other produce of their own farms, premises or manufactories situated within the boundaries of Colorado, and persons selling only Bibles and other religious publications shall not be deemed peddlers or hawkers within the meaning of this ordinance.

SHOWS AND EXHIBITIONS.

Sec. 12. Any person who shall own, conuct [sic] or manage for gain, within the limits of the town, any theater, circus, caravan, or any other exhibitions, show or amusement, or who shall exhibit any natural or artificial curiosities, or panoramic or other show device of any kind, or who shall give any concert or musical entertainment, without first having obtained a license therefor, shall forfeit and pay to the town a sum of not less than ten nor more than fifty dollars for each offense.

Provided, That for musical parties or concerts and exhibitions of paintings and statuary, given or made by citizens of the town for charitable or benevolent purposes, no license shall be required.

DATE—TIME.

Sec. 13. The mayor and recorder is [sic] authorized to issue a license to any person applying therefor for the purpose named in the foregoing section, upon the applicant paying the sum of from five to twenty-five dollars, at the discretion of the mayor and recorder, for each exhibition or performance, and such license shall specify the object, length of time and number of performances for which the same is issued.

DRAYS, EXPRESS, CABS, ETC.

Sec. 14. That, on and after the passage of this ordinance it shall be unlawful for any person to carry on or follow the trade, calling or business, for hire, of an expressman, cabman, drayman or public carrier within the limits of the town of Bachelor, without first having obtained a license therefor, as hereinafter provided.

CHARGES LIMITED.

Sec. 15. Any person who shall desire to carry on any such business as referred to in section 14 of this ordinance shall first procure from the mayor and recorder, who are authorized to issue the same upon payment of three dollars for each quarter of a year or twelve dollars for each year, a license for each and every vehicle used by such person in carrying on such business, said license to be issued for a period of not less than three months or more than one year; nor shall such expressman, drayman, cabman, or public carrier charge more than twenty-five cents for carrying anything of two hundred pounds weight or under, nor more than twenty-five cents per hundred for carrying each additional hundred weight or fraction thereof, within the limits of the town of Bachelor.

Sec. 16. Any person who shall violate this ordinance shall, upon conviction thereof, be liable to a fine of not less than three dollars nor more than twenty-five dollars, to be collected by law.

PUTTING UP LICENSE.

Sec. 17. Every person doing business at any fixed or permanent place in the town who is required to take out and have a license for such business, shall keep his license posted up in some conspicuous place in his place of business so as to be at all times exposed to view.

Sec. 18. Any person violating the provisions of section 17 of this ordinance shall be subject to and pay a fine of not exceeding five dollars and a further fine of five dollars for each day said person shall fail to comply with the requirements of said section.

Passed by the board of trustees and approved by me this 19th day of July, AD, 1892.
GUSTAV HOFFMAN, Mayor.
Attest: T. W. VINCENT, Recorder.[5]
[SEAL.]

SERIES OF 1892, NO. 4.

An ordinance relating to police magistrate, defining his powers, duties, etc.

Section 1. The board of trustees shall appoint, at the time of appointing other town officers, or as soon thereafter as practicable, a qualified elector of the town as police magistrate, who shall hold his office for one year and until his successor shall be duly appointed and qualified, unless sooner removed by the board of trustees.

Sec. 2. The police magistrate shall have jurisdiction to hear and determine all suits and prosecutions for the violation of any of the ordinances of the town of Bachelor.

Sec. 3. The police magistrate shall not issue a summons or a warrant for the arrest of any person without having at the same time a complaint, affidavit or statement, made under oath, charging the violation of some ordinance. The mayor and each of the trustees as well as the marshal and his assistants, are authorized to make arrests without a warrant where the offense is committed in view, but before the prisoner shall be put on trial, a statement, complaint or affidavit shall be made under oath specifying the charges against the prisoner.

Sec. 4. No action before a magistrate shall be dismissed for any defect in form in said statement, complaint or affidavit, if it substantially sets for the nature of the violation alleged, so as to give the defendant notice of the charge he is required to answer.

Sec. 5. Upon the filing of such statement or complaint, the magistrate shall enter the case upon his docket in the usual manner required by law, and shall issue a summons or warrant returnable forthwith, or at such time as the magistrate may designate.

Sec. 6. Any person arrested by virtue of a warrant, as hereinbefore provided, may be admitted to bail by executing a bond in double the amount of the penalty for the offense charged, conditioned that he will appear on a day mentioned before the magistrate, and not depart the court without leave; said bond shall have one or more sureties, unless when the magistrate is satisfied the defendant is a resident property owner, he may give his personal bond. All bonds shall be attested by the magistrate and filed, and an entry of the filing thereof shall be made in his docket.

Sec. 7. If the defendant fails to appear according to the condition of the bond, or appearing shall depart the court without leave, the magistrate shall enter judgment against him and his sureties for the penalty of said bond.

Sec. 8. Upon entering such judgment, the police magistrate shall issue a process in the name of the people of the state of Colorado against the parties liable upon such bond, requiring them to appear on the day mentioned before him and show cause why judgment should not be confirmed against them and execution issued. And such judgment may be set aside by the magistrate upon said defendant or sureties appearing and showing good cause for the non-appearance of the principal in such bond.

Sec. 9. A party in custody who can not be tried on account of the absence of witnesses or other cause, and who can not give bail for his or her appearance, may be confined in the town or county jail, or other places of confinement provided for the purpose, not exceeding three days and in such case the police mogistrate [sic] shall deliver to the officer committing such person a commitment stating the cause of such detention.

Sec. 10. When a defendant duly summoned fails to appear at the time set for the trial, the magistrate shall hear and examine the witnesses offered on the part of the town and shall render judgment by default against the defendant for such sum as the magistrae [sic] may under ordinance deem proper.

Sec. 11. Upon the rendition of judgment against the defendant for violating any ordinance of the town, the magistrate shall make an order and enter the same upon his docket, that if the defendant refuse or neglect to satisfy such judgment and costs of suit, he shall be confined in the town jail or county jail, or such other place as the board of trustees may from time to time designate. Execution may be issued on the rendition of judgment and placed in the hands of the town marshal for collection.

Sec. 12. Each person against whom any penalty, fine or forfeiture shall be recovered under any ordinance of the town who shall refuse or neglect to pay the same when demanded upon execution, shall be committed to the town or county jail, or such other place as the board of trustees may designate, and shall labor at such work, within the town, as his or her strength will permit until said fine, penalty or forfeiture and all costs thereon are fully paid, to be allowed at the rate of one dollar per day for each day's work performed under the direction of the street commissioner or marshal; provided that no such person shall be required to work more than eight hours a day; and provided further, that no such imprisonment shall exceed ninety days for any one offense.

Sec. 13. The town marshal at the first regular monthly meeting of the board shall make a report to the board of trustees of the number of days' work performed in pursuance of this ordinance and by whom performed.

Sec. 14. In all prosecutions instituted by the town of Bachelor, any officer or citizen shall be a competent witness for the town.

Sec. 15. The police magistrate shall keep a record of all cases tried and determined by him in a well bound record book, which book shall be the property of the town and shall be transmitted to his successor in office. The method of procedure in his office shall conform as nearly as may be to the practice before the justices of peace in this state.

The magistrate shall be entitled to such compensation for his services in fees as by statute are allowed to justices of the peace in counties of the fifth class for like services.

The magistrate shall tax as costs such attorneys fees as may hereafter be agreed upon by the board, in each case prosecuted by the town attorney, which shall be collected as other costs in the case and which shall be paid into the town treasury. Provided, however, that in no case shall the town of Bachelor be liable for any costs incurred in the police magistrate's court.

Sec. 16. It shall be the duty of the police magistrate to collect all fines, penalties and forfeitures and pay the same over to the town treasurer on or before the first day of each month, and he shall also make a report of the number and amounts of fines, on what account, and of whom uncollected and the reason thereof, to the board of trustees on the first regular meeting of each month, said report shall embrace the doings of the police magistrate for the month prior to making such report. Whenever said magistrate shall pay any money into the town treasury, he shall take duplicate receipts therefor, one of which shall accompany his monthly report to the board of trustees, and the other said magistrate shall keep.

Sec. 17. In all prosecutions for fine or penalty, when the defendant shall be acquitted, the informer or prosecutor may, in the discretion of the magistrate, be adjudged to pay the costs, if it appear[s] to the satisfaction of the magistrate that the prosecution was instigated vexatiously or without reasonable cause.

Sec. 18. Any complaint made for the violation of an ordinance of the town of Bachelor may be made upon information and belief. There shall be no change of venue granted in any case.

Sec. 19. The police magistrate before entering upon the duties of his office, shall execute a bond to the town of Bachelor in the penal sum of one thousand dollars, conditioned upon the faithful and impartial discharge of the duties of his office, that he will promptly pay all fines, forfeitures and penalties to the town treasurer as required by this ordinance, and will promptly deliver to his successor in office all books, papers and other property belonging to the town and appertaining to his office.

Sec. 20. The mayor shall be authorized, whenever the occasion may seem to warrant, to remit any fine, forfeiture or penalty imposed by the police magistrate for the violation of any ordinance of the town, upon payment of costs by the offender in any such case, which remittance shall be in writing and signed by the mayor, and to be delivered to the police magistrate and by said magistrate to be noted upon the docket, and such remittance shall be noted in the magistrate's monthly report to the board of trustees.

Passed and approved this 19th day of July, 1892.

GUSTAV HOFFMAN, Mayor.

Attest: T. W. VINCENT, Recorder.[6]

[SEAL.]

SERIES OF 1892, NO. 5.

An ordinance concerning nuisances.

Be it ordained by the board of trustees of the Town of Bachelor:

Section 1. Whoever shall deposit, throw or discharge or leave any offensive, foul or putrid liquor, substance or excrement, or any liquor or substance likely to become nauseous, foul, offensive or putrid, within the town, or so as to be or likely to become injurious to the health or comfort of any person residing within the town, shall be guilty of a nuisance and subject to a penalty of from $10 to $25, and a like penalty for each day he shall not abate or remove such nuisance after notice.

Sec. 2. Whoever shall knowingly suffer or permit any dead animal belonging to him to remain within the town, so as to be or likely to become putrid or nauseous or offensive to any person residing within the town, shall be deemed guilty of a nuisance and be subject to a penalty of not less than $10 and not exceeding $25.

Sec. 3. When any nuisance or anything likely to become a nuisance shall be found by the street supervisor, town marshal, or member of the board of health, or shall be reported to them, the owner, author or cause of such nuisance shall forthwith be notified to abate, remove or remedy the same; and in case he shall not comply with such notice, the officer shall abate such nuisance and bring suit against such person in the name of the people of the state of Colorado, in behalf of the town, for the penalty and the costs of removing the same, which may be recovered with the penalty, or in a separate suit for the same, before any court having jurisdiction.

When any nuisance or anything likely to become a nuisance may be found upon any premises, and the owner, author or cause of such nuisance is unknown or can not be found, the owner, occupant or agent of such premises shall be notified to abate or remove the same, and if such owner or his agent or occupant shall not comply with such notice, he shall be subject to a penalty of from $10 to $25 for each day he shall not abate the same, and the officer may abate the same and may bring suit in the name of the people of the state of Colorado, on behalf of the town, against the person liable for the penalty and costs of removal or abatement, as in other cases; or, if no person liable therefor can be found, may report the costs, under oath, to the town board for allowance.

When the owner, author or cause of any nuisance, or the owner or his agent or the occupant where such nuisance exists is unknown, or can not be found within the town, the officer shall abate such nuisance forthwith without notice and shall report the cost to the board for allowance.

Sec. 4. Whoever shall throw or deposit in or near any waters furnishing water for domestic use to the people of the town of Bachelor any dead animal, putrid, foul or poisonous substance, or anything likely to become putrid, foul or poisonous shall be deemed guilty of a nuisance and shall be fined in any sum not exceeding $25.

Sec. 5. Any person who shall permit any cellar, vault, privy, drain, pool, sewer or grounds upon any premises belonging to or occupied by him to become nauseous, foul or injurious to the public health shall be subject to a fine of not less than $10 (ten dollars) for each day that the same is suffered to remain after notice by the board of health to abate such nuisance.

Passed by the board of trustees and approved by me this 19th day of July, AD 1892.
GUSTAV HOFFMAN, Mayor.
Attest: T. W. VINCENT, Recorder.[7]
[SEAL.]

SERIES OF 1892, NO. 6.

An ordinance defining the town seal.
Be it ordained by the board of trustees of the town of Bachelor:

That the seal of the town of Bachelor shall be one and five-sixths inches in diameter, with the following device inscribed thereon: Two circles, one within the other, between which shall be the words "The Town of Bachelor, Colorado," and besides these words there shall be two small decorative devices. Within the inner circle shall be the word "Seal," with small decorative devices above and below said word.

Passed by the board of trustees, and approved by me this 19th day of July, AD 1892.
GUSTAV HOFFMAN, Mayor.
Attest: T. W. VINCENT, Recorder.[8]
[SEAL.]

Fig. 74. Embossed town seal of Bachelor on paper, made with the tool shown in Fig. 23. Courtesy of the Grant Houston and the Hinsdale County Museum, Lake City, Colorado. Photograph by the author, 2016.

SERIES OF 1892, NO. 7.

An ordinance to prohibit the carrying of concealed weapons and provide punishment therefor.

Be it ordained by the board of trustees of the town of Bachelor:

Section 1. If any person shall, within the corporate limits of the town of Bachelor, carry concealed upon or about his or her person any pistol, revolver, bowie knife, dagger or any other deadly weapon, such person shall, upon conviction thereof, be punished by a fine of not less than twenty-five ($25) dollars, or imprisoned in the town jail for a term of not less than ten nor more than sixty days, or both, in the discretion of the court. Provided, That this section shall not be construed to sheriffs or other officers of the peace while on duty.

Sec. 2. If any person shall have on his or her person, within the corporate limits of the town of Bachelor, any pistol, revolver, gun, knife, dirk, bludgeon or other deadly weapon with intent to assault any person, every such person, on conviction shall be fined in any sum not exceeding five hundred dollars, or imprisoned in the town jail not exceeding three months.

Sec. 3. The police magistrate of the town of Bachelor shall have final jurisdiction in such cases, subject to appeal as provided by law.

Passed by the board of trustees, and approved by me this 19th day of July, AD 1892.

GUSTAV HOFFMAN, Mayor.

Attest: T. W. VINCENT, Recorder.[9]

[SEAL.]

SERIES OF 1892, NO. 8.

An ordinance concerning a road poll tax, and provisions for collection of same.

Be it ordained by the board of trustees of the town of Bachelor.

Section 1. Each and every able bodied male citizen within the corporate limits of the town of Bachelor shall annually pay to said town the sum of three dollars as a poll road tax, or in lieu thereof perform two days' labor on the roads within the corporate limits of said town, under the direction and supervision of the town marshal while acting as street commissioner.

Sec. 2. Should any such person, after ten days' written notice, to be served on him by the town marshal, to pay said three dollars or do said two days' work has been given to him, refuse, fail or neglect to pay said sum or do said work, the town marshal shall cause suit to be entered against said delinquent person before the police magistrate for said tax and costs.

Sec. 3. If said poll road tax be collected in cash, the same shall be paid to the town treasurer and placed to the credit of the street improvement and bridge fund of said town.

Sec. 4. The town marshal shall give to each person paying said poll tax in cash or work a receipt for the same, and at the end of each month said marshal shall report to the board of trustees the names of the parties so paying or working.

Passed by the board of trustees, and approved by me this 19th day of July, AD 1892.

GUSTAV HOFFMAN, Mayor.

Attest: T. W. VINCENT, Recorder.[10]

SERIES OF 1892, NO. 9.

An ordinance concerning appropriations for the fiscal year beginning July 19th, AD 1892, and ending July 18th, AD 1893.

Be it ordained by the board of trustees of the town of Bachelor:

Section 1. That for the purpose of defraying the necessary expenses and liabilities of the said town of Bachelor for the fiscal year beginning May 1st, AD 1892, and ending April 1st, AD 1893, there shall and hereby is appropriated several sums of money hereinafter set forth for the following objects and purposes, to wit:

For salaries of officers of said town, the sum of three thousand dollars.

For the construction, improvement, maintenance and repair of streets, alleys, bridges and waterways, the sum of three thousand dollars.

For the purchase of furniture, rent of town hall, calaboose and other contingent expenses, the sum of three thousand dollars.

Sec. 2. The town treasurer of the town of Bachelor, except as to poll road tax, shall place to the credit of the salary fund, the improvement fund and the contingent fund such proportions of the amounts received by him to each of the above funds as the amount appropriated to each fund bears to the whole amount appropriated.

Passed by the board of trustees, and approved by me this 19th day of July, AD 1892.

GUSTAV HOFFMAN, Mayor.

Attest: T. W. VINCENT, Recorder.[11]

[SEAL.]

SERIES OF 1892, NO. 10.

An ordinance concerning fire—prevention thereof.

Be it ordained by the board of trustees of the town of Bachelor:

Section 1. That the town marshal shall be and is hereby made the fire warden of the town of Bachelor.

Sec. 2. It shall be the duty of the fire warden to see that ordinances of the town concerning the fire department are enforced.

Sec. 3. It shall be the duty of the fire warden to examine all buildings, whether occupied or not; all depositories of ashes and manufacturing establishments, and report to the board of trustees any violation of this ordinance in relation thereto, as well as on the subject of prevention and extinguishment of fires within the town.

Sec. 4. The fire warden shall from time to time inspect all stoves, fireplaces, hearths, grates, furnaces, boilers, and other places in the town where fires may be kept, and all stove-pipes, chimneys, ovens and other apparatus or fixtures connected therewith for the purposes of ascertaining if the same be in a safe and proper condition; and, whenever, in the judgment of the fire warden, any such place for keeping fire or for conducting heat or smoke shall be in such condition as to render the keeping of fire unsafe therein, or the conducting of heat or smoke unsafe, it shall be his duty to order the occupant of the building to which the same may be attached to discontinue the making of fires therein, and to make all necessary repairs to render the same safe when fire is used in such building, in the opinion of the fire warden inspecting the same; and any person who shall disobey such order, when the same is given in writing by the fire warden, shall forfeit and pay to the town the sum of not less than $5 and not more than $10 for each and every day he shall continue to use the same in disobedience to such order, to be recovered as in actions for debt.

Provided, That any person may appeal to the board of trustees who feels himself aggrieved by any such order of the fire warden, but such appeal shall not operate to suspend such order further than such part as may concern alterations or repairs, which shall be suspended until the final decision of the board of trustees on such appeal. Any person who shall resist or hinder the fire warden in the execution of any duty herein enjoined shall forfeit and pay to the town of Bachelor not less than $5 nor more than $10 for each offense.

Sec. 5. No lighted candle or lamp shall be used in any stable or other place wherein hay, straw or other combustible material is kept unless such lamp or candle be well secured in a lantern, under penalty of not less than 25 and not more than 50 dollars for each offense.

Sec. 6. No hay, straw, chips or other combustible material shall be set on fire or burned in any street or place within 100 feet of any building within the town without the permission of the fire warden, under penalty of not less than ten and not more than twenty dollars for each offense.

Sec. 7. No ashes shall be kept or deposited in any part of the town unless the same be kept and deposited in a secure place made of brick, stone, metal, or any other incombustible material. Any person violating any provision made in this ordinance shall be liable to a fine of not less than five and not more than ten dollars for each offense.

Sec. 8. Any person who shall within ten feet of any building wherein fire is kept within the town place any hay, straw, or other combustible material, in sack or pile, without having the same enclosed or secured so as to protect it from flying sparks of fire, shall be subject to a fine of five dollars for each offense and a fine of ten dollars for each week the same shall be allowed to so remain after notice by the fire warden for its removal.

Sec. 9. Any person who shall set up or use any stove without having under it a sheet of tin, zinc or other incombustible material, or who shall place or use a stove the upper part within two feet of any wood work, without protecting such wood work with metalic [sic] covering, so as to prevent the same from taking fire, shall forfeit and pay to the town of Bachelor a sum of five dollars and a further sum of ten dollars for each week the same shall so remain after notice by the fire warden.

Sec. 10. No person shall keep at his place of business a greater quantity of gunpowder or other explosive material than 50 pounds at one time, and the same shall be kept in tin or copper canisters or cases containing not more than ten pounds each, and in a place remote from fires, lighted lamps or candles, and no person shall weigh or expose for sale any gunpowder, gun cotton or other explosive material after the lighting of lamps, unless the same be in sealed canisters or cases, nor shall any person sell or deliver any coal oil, nitro-glycerine or other like combustible fluid after lamplight. A violation of any provision of this ordinance shall subject the offender to a fine of not less than 25 dollars nor more than 100 dollars for each offense.

Passed by the board of trustees, and approved by me this 19th day of July, AD 1892.

GUSTAV HOFFMAN, Mayor.

Attest: T. W. VINCENT, Recorder.[12]

[SEAL.]

SERIES OF 1892, NO. 11.

An ordinance concerning offenses against the public peace and to provide the punishment thereof.

Be it ordained by the board of trustees of the town of Bachelor:

Section 1. If any person maliciously or willfully disturb the peace or quiet of any neighborhood or family in the town of Bachelor by loud or unusual noises or by tumultuous or offensive carriage, threatening, traducing, quarreling, challenging to fight, or fighting or making himself or herself obnoxious by any of said things in any of the streets or public places, in said town, he or she shall, on conviction, be fined in a sum not less than five dollars nor more than fifty dollars, or imprisoned in the town calaboose not exceeding thirty days.

Sec. 2. If two or more persons assemble in said town for the purpose of disturbing the public peace or committing any unlawful act, and do not disperse on being desired or commanded so to do by any peace officer, such persons so offending shall, on conviction, be severally fined in any sum not exceeding fifty dollars.

Sec. 3. Any person who shall become intoxicated in any of the streets or public places in said town, or shall use vulgar or profane language in such places, shall, on conviction, be fined in a sum not to exceed fifty dollars.

Sec. 4. If any person or persons shall discharge any firearms either by day or night for the purpose of disturbing the public peace, or without good reason for so doing, on conviction, may be fined a sum not exceeding $50.

Passed by the board of trustees, and approved by me this 19th day of July, AD 1892.

GUSTAV HOFFMAN, Mayor.

Attest: T. W. VINCENT, Recorder.[13]

[SEAL.]

SERIES OF 1892, No. 12.

An ordinance to regulate the speed of driving or riding animals and to provide for the enforcement of the same.

Be it ordained by the board of trustees:

Section 1. No person shall ride any animals or drive any animals attached to any vehicles in the town of Bachelor in any of the streets, alleys or public highways therein at a speed exceeding six miles per hour.

Any one violating the provisions of this ordinance, on conviction, shall be fined in the sum of not less than one dollar nor more than twenty-five dollars.

Passed by the board of trustees, and approved by me this 19th day of July, AD 1892.

GUSTAV HOFFMAN, Mayor.

Attest: T. W. VINCENT, Recorder.[14]

[SEAL.]

SERIES OF 1892, No. 13.

An ordinance establishing the office of town physician.

Be it ordained by the board of trustees of the town of Bachelor:

Section 1. The office of town physician is hereby established, and the board of trustees shall annually hereafter appoint a competent physician to fill said office and shall fix by resolution or contract the compensation for his services.

Passed by the board of trustees, and approved by me this 19th day of July, AD 1892.

GUSTAV HOFFMAN, Mayor.

Attest: T. W. VINCENT, Recorder.[15]

[SEAL.]

SERIES OF 1892, No. 14.

Be it ordained by the board of trustees:

Section 1. The mayor, board of trustees, the town physician and the town marshal be, and they are hereby constituted a board of health for the town of Bachelor.

Sec. 2. The said board of health shall have power, and it is hereby made their duty, to take all steps necessary to prevent the spread of infections or contagious diseases within said town, and in the discharge of said duty may remove by force, if actually required, any person or persons so diseased; they may purchase or rent houses for hospitals, hire nurses and incur any and all necessary expense in the removal and treatment of such diseased persons, and such actual and necessary expense shall be audited and allowed by the board of trustees in case the patient or patients is or are unable to pay the same.

Sec. 3. Said board of health shall have power, and it is hereby made their duty to order any nuisance found by them injurious to the health of the town abated, or removed without the corporate limits of the town.

Sec. 4. All orders issued by the said board of health shall be in writing and signed by the mayor, and shall be executed by the town marshal.

Sec. 5. Any person failing to comply with any orders made by the said board of health as in this ordinance directed shall, upon conviction, be fined not less than three nor more than one hundred dollars, and shall remain in custody until such fine and costs are paid, not exceeding sixty days.

Passed by the board of trustees, and approved by me this 19th day of July, AD 1892.

GUSTAV HOFFMAN, Mayor.

Attest: T. W. Vincent, Recorder.[16]

[SEAL.]

SERIES OF 1892, No. 15.

An ordinance concerning accounts.

Be it ordained by the board of trustees of the town of Bachelor:

Section 1. No account or claim against the town of Bachelor, except for the salaries of its officers, or for the payment of a special contract made by the board of trustees or by some officer of the town authorized by them to make the same, shall be audited or allowed unless the same shall be itemized and sworn to, and no account or claim except those above mentioned shall be allowed at the meeting of the board of trustees to which the same is first presented, unless by consent of all the trustees present.

Passed by the board of trustees, and approved by me this 19th day of July, AD 1892.

GUSTAV HOFFMAN, Mayor.

Attest: T. W. Vincent, Recorder.[16]

[SEAL.]

SERIES OF 1892, No. 15.

An Ordinance concerning Dogs.

Be it ordained by the City Council of the town of Bachelor:

Section 1. Every owner or possessor and every person who harbors or keeps any dog within the corporated [sic] limits of the Town of Bachelor, shall annually pay to the city treasurer for the use and benefit of [the] city, the sum of one dollar for each and every male dog, and the sum of two dollars for each and every female dog and whelps owned, possessed, harbored or kept by any such person.

Sec. 2. On the payment of said license fee to the treasurer, the receipt thereof being produced to the city clerk, the said clerk shall, in order of application, register and number the name of the licensee and deliver to him a duly executed license, which shall be numbered and state name of person taking out the same, and the clerk shall also deliver to each person a metallic check or tag, numbered with a number corresponding with that of the registry and license; such license and registry shall entitle the licensee to keep said dog until the 2nd day of Aug. next ensuing, but no longer.

Sec. 3. No owner or keeper of any dog or bitch shall allow or permit such dog or bitch to be at any place within this city other than upon his or her own premises at any time without a collar having attached thereto the metallic check or tag hereinbefore referred to; nor shall any owner or keeper of any dog or bitch permit or allow the same to wear any check or tag other than the identical one issued for such dog or bitch. In case of loss a duplicate check or tag shall be issued by the city clerk at the expense of the person making the application.

Sec. 4. No person shall remove or cause to be removed the collar, check or tag from any registered dog or bitch, without the consent of the owner or keeper thereof.

Sec. 5. No person shall keep, own, or harbor any dog or bitch which is notoriously vicious or dangerous, nor one which by long and frequent or habitual barking, howling or yelling shall annoy or disturbe [sic] any neighbor hood [sic], or which shall run after, bark at, frighten, molest or disturb any horse, mule or working animal driven by or in charge or [sic] any person.

Sec. 5 [6]. It shall be the duty of the city marshal and policemen to kill any bitch running at large within corporate limits, while in heat, and it shall be the duty of the city marshal and policemen to kill any dog found running at large within the city limits, for which no license fee has been paid. All dogs killed by the city marshal shall be removed beyond the city limits and buried.

Sec. 7. Any person convicted of violating any of the provisions of this ordinance shall be fined in the sum of not less than five dollars or more than fifty dollars.

Adopted and passed this 2nd day of August, 1892

GUSTAV HOFFMAN, Mayor.

Attested by the undersigned with the corporate seal of the Town of Bachelor.

T. W. VINCENT, City Clerk.[18]

Appendix 2
Colorado Business Directory
Entries for Bachelor

These listings are reproduced largely as they appeared in the respective business directories. The text is verbatim; the main difference is that the longer listings are on a single line to save space. Because the business directories used a two-column format, the longer entries were on two or three lines in the original directory. A few listings out of alphabetical order in the originals were placed in alphabetical order. I have retained entries that were in boldface in the original directories. Entries listed on a single line in the original were frequently abbreviated, and that feature has been retained as well.

1893[1]
BACHELOR (Teller, P. O.)

Bachelor, Creede Camp, by virtue of its more favored location, is to-day the business center of Creede Camp. Beautifully situated in two of those famous mountain parks, each park being a step—one about fifty feet higher than the other—10,100 feet above the level of the sea, or about the same elevation of Leadville, which is 10,220 feet. The view from this town, over to the southern range and up the Rio Grande River for 100 miles is declared by travelers to be one of the finest on this continent.

Bachelor City is one of those happy combinations of chance and necessity so seldom to be found together when fate determines that a city shall be born. When that most remarkable body of ore that has ever been discovered, was opened to the world showing the necessity for the employment of thousands of men, and the transaction of millions of business, there, right at hand were those beautiful parks with their sparkling springs, ready and inviting, and Bachelor City sprang into existence as if by magic. The town site was surveyed in January, 1892, and in July of the same year the city was incorporated, with a full set of officials. The two parks, covering some eighty acres of ground, were covered with houses of all descriptions occupied by people of nearly every nationality. As this time, February, 1893, Bachelor is doing two-thirds of the business of Creede Camp. The post-office receipts indicating [sic] a population of about 6,000 people.

Bachelor has several first class hotels, a bank, several livery stables, blacksmith shops, hardware stores, grocery stores, confectioners, dry goods stores, bakeries, churches, Sunday-schools, dance-houses, a first class district school, saloons galore, lawyers, notaries, justices, woodhaulers, water venders [sic], newspapers and newsboys, bootblacks, and you can buy wieneworst [sic] and hot tamales [sic] at your door anytime. Bachelor has a miners' union organization in full blast and several of the secret organizations are represented. For so promiscuous a crowd, gotten together in so short a time, the town is comparatively quiet. It is almost useless to say that the voters of Bachelor cast a vote almost unanimous for the only silver candidate.

If anyone asks if this a mushroon [*sic*], or has Bachelor the elements of permanency, we only have to point to the results of the first year's development of the mines right at our doors, and when one remembers that the first year's work in a mining camp is work of development and preparation, the result is the most phenominal [*sic*] ever shown by any camp on earth—not excepting Leadville in its first year. The statistics in brief for Creede camp in its first year was [*sic*] that it showed up fifteen pay mines, there are twenty now. The number of cars of ore shipped the first year was 3,517 cars; tons, 46,355. Value five millions [*sic*] dollars. This output has increased every day until the result has startled every mining camp in America. We have the result for February, 1893, 9,000 tons amounting to $800,000, and all this from what is known as superficial development. Can the human mind conceive what the future of Bachelor City shall be when this vast body of mineral shall have been developed and it lays right at our doors.

Armstrong, Miss Jennie, lndy.
Bachelor City Hotel, Mrs M H Maxwell, prop.
Bachelor Grocery Co, Frank Gillett, mgr, wh&retgrocrs.
Bachelor Transportation Co, L J Chapman, manager.
Biles, J A, physician.
Bishop, Mrs Maggie, prop City Hotel.
Bruner, Rosa L, millinery and notions.
Carey & Lamb, saloon
Chapman, L J, blacksmith, transfer and livery.
City Barber Shop and Baths, E J Marsell, prop.
City Council, Geo Martindale, T A Armstrong, C A Jones, A H Whitehead, John Gould and W W Ellsberry.
City Hotel, Mrs Maggie Bishop, prop.
Coffin, S D, mine prospector.
Collins, P J, saloon.
Commercial House, G W Stone, proprietor.
Covert, E C, general mdse.
Cunningham & Sloan, lumbr.
Dougherty, J H saloon.
Ellsberry & Foutch Banking and Mercantile Co.
Free Coinage Hotel, Miller & Mitchell, props.
Gair, C J, attorney.
Gavin, C J, attorney.
Gillett, Frank, manager Bachelor Grocery Co.
Gould, John, postmaster.
Grundy J F, saloon.
Gunn, A, saloon.
Hankey, A, laundry.
Harvey & Holt, boa'd'g house.
Hilton, E F, photographer.
Hoffman, Gus, mayor, hardware and notions.
Hudlow, J W, 2nd hand goods.
Jones & McCulloch, boots and shoes.
Karrick, Joe, mine owner.
Leary, T F, city marshal.
Lee & Kunz, barbers.

Links, F L, jeweler.
Lockwood, L E, bakery.
Love, Ella C, stationery and news stand, post-office.
Lucette & Co, R J, hardware.
Lundy & Sherry, saloon.
McDonald, Mrs Agnes, boarding house.
McDonald, A J, saloon.
McDougall & Smith, meat market.
McGregor, A, carpt & buldr.
McKenzie J, merch'nt tail'r.
McLean, merchant tailor.
Marsell, E J, prop city barber shop and baths.
Marshall, Mrs M H, prop Bachelor City hotel.
McManus, Thomas, stationery, groceries, cigars & notions
Merritt, Wm H, wall paper, paints, etc, painter.
Miller & Mitchell, props Free Coinage Hotel.
Myers & Co, H, clothing.
Parker, T E, meat market.
Roe, C W, wall paper, paints, etc.
Rosen, F C, saloon.
Shorten, John, Ed *Teller Topics*.
Stone, G W, prop Commercial House.
Tam & Lang, bakery and restaurant.
Teller House, A H Whitehead, proprietor.
Teller Topics, (w) John Shorten, publisher.
Van Noorden, S E, attorney and justice peace.
Van Norman, physician.
Vincent, T W, lumber, coal, real estate, town clerk and treasurer.
Wells, Mrs E D, dry goods.
Whitehead, A H, prop Teller House.
Williams, J B, saloon.
Willis & Glenn, groceries, etc.
Willoughby & Collar, clothing, shoes and hats.

1894²

BACHELOR—(Teller, P. O.)

The center of the miner population of Creede camp, two and one-half miles northwest of Jimtown (City of Creede) and a half mile from the producing mines. Population 800. Altitude 10,100 above the level of the sea. It is an incorporated town with a full quota of officials; churches, its own school district, all lines of commercial pursuits, and destined to be the liveliest part of the great silver producing [*sic*] section of Colorado.

BACHELOR GROCERY CO., F. Gillette, mgr., wholesale and retail groceries.
Belcher, A. C., blacksmith and wagon-maker.
Biles, J. A., physician.
Bishop, F. F., prop. Last Chance Boarding house.
Broade, E. E., sawmill and lumber.

Brown, Geo. T., clerk of council.
BROWN, LOU B., clairvoyant.
Bruner, R. L., bakery and confectionery.
Calverd, M. R., barber.
CAREY & REGAN, props. Fashion saloon.
Cassidy, Dollie, teacher public school.
Coffin, S. D., mine prospector.
Collins, P. J., prop. White House saloon.
Covert, E. C. groceries.
Crawford & McDonald, meat market.
Crawford, E. H., supt. N.Y. and Last Chance mines.
CUNNINGHAM & SLOAN (J. B. Cunningham and W. C. Sloan), sawmill and lumber.
DOERING, A. H., justice peace.
Doherty, J. H. saloon.
Eades, A. B., dry goods.
Eckel, Rev. F. E., pastor Congregational Church.
Fay, A. C., pres. Western Telegraph Co.
Gould, Jno., postmaster.
Hankey, A., laundry.
Henderson, J. D., woodyard.
Henneberg, Emma M., millinery and dressmaking.
Jenkins, Jos. W., Gem saloon.
Jones & McCulloch, boots, shoes and furnishings.
Karrick, Jos., mine owner.
Leary, T. F., engineer N. Y. C. Mine.
Likins, F. L., drugs and jewelry.
Lucette, R. J. & Co., hardware.
Mallon, Mrs. Mary, saloon and boarding house.
McCarty, Mat, saloon.
McDonald & Jones, Palace Club rooms.
McDOUGALL & SMITH, meat market.
McKenzie, J., merchant, tailor.
McLean, A. merchant tailor.
McManus, Thos., stationery, confectionery, notions, cigars and tobacco.
Miller, A. L., mail and express.
Page, J. B., assayer N. Y. Chance Mine.
Parker, T. E., prop. Park boarding house.
Parkwell, Phillip, barber.
Reynolds, F. B., prin. Public school.
Ridenhour & Brooks, livery.
Roe, C. W., painter and paperhanger.
Shorten, Jno., editor *Teller Topics*.
Swords, Mrs. M., prop. Bachelor City Hotel.
Teller House, A. H. Whitehead, prop.
Teller Topics (w), John Shorten, editor.
Tognana, Mrs. Katie, laundry.
Van Noorden, S. E., attorney, town attorney, county supt. schools.
Vincent, T. W., real estate, coal, building materials and town treasurer.
Warren & Coulsen, furniture.

Weinrich, M., cabinet maker.

Wells, E. B., general merchandise.

Willians, J. H., restaurant and bakery.

Williams, W. W., marshal.

1895[3]

BACHELOR (Teller, P. O.)

The center of the miner population of Creede camp, two and one-half miles northwest of Creede. Population 800. Altitude 10,100.

Bachelor Grocery Co., F. Gillett, mgr.

Barnett, H. M. architect and builder.

Barnie and Rice, groceries and meat market.

Belcher, A. C., blacksmith.

Biles, J. A. physician.

Bishop, F. F. prop. Last Chance boarding house.

Broade, E. E. saw mill and lumber.

Brown, Lou B. clairvoyant.

Bruner, Mrs. Rosa L. millinery and dressmaker.

Calvert and Fulst, barbers.

Carey, S. baths.

Charlton, Mrs. W., boarding.

Charlton, Wm. engineer Amethyst mine.

Colville and McKenzie, gents. furnishings.

Comestock, Mrs. M. J., Boarding.

Crawford, E. H. supt. New York Chance Mine.

Cummings, Agnes, teacher public school.

Cunningham and Sloan, saw mill and lumber.

Doering, A. H., justice peace.

Doherty, J. H., Saloon.

Eades, A. B., dry goods.

Henneberg, Emma and Minnie, dressmakers.

Hughes, W. M., blacksmith.

Jenkins, J. W., Saloon.

Jones, A., boots, shoes and furnishings.

Lee, C. B., barber.

Likins, F. L., drugs and jewelry.

Loucks, W., transfer.

MacLean, A., mer. Tailor.

Manion, Mrs. M., boarding.

Markley, Tailor, blacksmith New York Chance Mine.

McDonald, A. J., Saloon.

McDougall, A. meat market.

McKenzie, A., engineer New York Chance Mine.

McKenzie, J., mer. Tailor.

McLeod, R. J., blacksmith New York Chance Mine.

McPhee, Angus, saloon.

McQuaig, Neil, engineer New York Chance Mine.

Miller, A. L., coal, feed and livery.

Miller, C. F., restaurant.
O'Leary, T. F., engineer New York Chance Mine.
Page, J. B., assayer New York Chance Mine.
Parker, T. E., Park Boarding House.
Pollock, Mrs. Laura J., principal public school.
Regan, M., saloon.
Ridenour & Henry, livery.
Rodman, Julius, resident agent, New York Chance Mine.
Rosen, Fred, marshal.
Sentinel, The, (w), C. O. Sprenger, editor, A. M. Sprenger, pub.
Swords, Mrs. M., Bachelor City Hotel.
Terrell, J. K., clothing, boots, shoes and furnishings.
Thomas, Rev. C. M., pastor Congregational Church.
Tognani, Mrs. Kate, laundry.
Van Noorden, S. E. county supt. schools, attorney and notary public.
Vincent, T. W., real estate, coal, building material, town clerk and treasurer.
Vincent's Opera House, T. W. Vincent, prop.
Warren and Coulson, furniture.
Weinrich, W., cabinet maker.
Wells, E. D., constable and gen. mdse.
Westlake, J. W., master mechanic New York Chance Mine.

1896[4]

BACHELOR (Teller, P. O.)

The center of the miner population of Creede camp, two and one-half miles northwest of Creede. Population 800. Altitude 10,100.

Bachelor Grocery Co., F Gillett, manager.
BARNETT, H M, architect and builder.
Barnie & Rice, groceries and meat market.
Bay & Co., Geo. W S, Last Chance boarding house.
Belcher, A C, blacksmith.
Biles, J A, physician.
Broade, E. E., saw mill and lumber.
Bruner, Mrs. Rosa L, millinery and dressmaker.
Charlton, Mrs. W, boarding.
Colville, John B, gents furnishings.
Crawford, E H, supt. New York Chance Mine.
Cunning & Sloan, saw mill and lumber.
Doherty, J H, saloon.
Eades, A B, dry goods.
Heilscher, August, jeweler.
Henneberg, Misses, dressm'ks.
HUGHES, W M., blacksmith.
Irwin, Rev. John, pastor Congregational Church.
Jenkins, J W, saloon.
Lee, C B, barber.

Likins, F L, drugs ank [sic] jewelry.
Loucks, W, transfer.
Lynch, Geo., confectionery.
MacLean, A, merchant tailor.
Mallon, H P, saloon.
Manion, Mrs. M, boarding.
Markley. Tailor, blacksmith New York Chance Mine.
MAY, I S, gen. mdse. and meat market.
McDonald, A K, blacksmith.
McDonald, Dan, marshal.
McKenzie, D B, mch't tailor.
McLeod, R J, blacksmith New York Chance Mine.
McPhee, Angus, saloon.
O K Clothing House, A C Monday, manager.
Page, J B, assayer New York Chance mine.
Pollock, Mrs. Laura J, teacher public school.
Ray, Ed M, supt Unibell Leasing Co.
Reilly, A F, town clerk and constable.
Ridenour & Henry, livery.
Rodman, Julius, resident agt. New York Chance mine.
Swords, Mrs. M, Bachelor City Hotel.
Terrell, J K, clothing, boots, shoes and furnishings.
VAN NORDEN, S E, post-master, att'y and notary.
VAN NORDEN, MRS. S E, news, confect'y & cigars.
Vincent, T W, coal & lumber.
Vincent's Opera House, Levi & Lewin, props.
Walsh, Mrs. John, boarding.
WARREN & COULSON, furniture.
Weinrich, W, cabinet maker.
Wells, E D, gen'l mdse.
Whitehead, J H, just. peace.

1897[5]

BACHELOR (Teller, P. O.)

The center of the mining population of Creede camp, two and one-half miles northwest of Creede. Population 300. Altitude 10,100.

Barnie, groceries and meat market.
Belcher, A C, blacksmith.
Biles, J A, physician.
BROADE, E E, saw mill and lumber.
CASEY, MARY J, postmistress.
Congregational Church, John Irwin, pastor.
Cunningham & Sloan, saw mill and lumber.
Doherty, J H, saloon.
Hannington, supt Last Chance mine.
Henneberg Misses, dressmkrs.
Jenkins, J W, saloon.

Jones, A, clothing.
Likins, F L, drugs & jeweler.
Lucette, R J, hardware.
Lynch, Geo, confectionary [*sic*].
Markley & Fagan, groceries.
McKenzie, D B, mercht tailor.
McPHEE, AUGUS, saloon.
Oliphant, F L, principal public schools.
Regan, Mike, saloon.
Russell, Mrs Ida, boarding.
Terrell, J K, clothing.
VINCENT, T W, coal & lumber.
Vincent's Opera House, Levin & Lewin, props.
Walsh, Mrs John, boarding.
Weinrich, W, cabinet maker.
Whitehead, J H, just peace.
Wyland, Geo A, meat market.

1898[6]

BACHELOR (Teller P.O.)

An important mining town in Mineral county, 2½ miles northwest of Creede. Population 200. Altitude 10,100.

Belcher, A C, blacksmith.
Biles, J A, physician.
Broade, E E, lumber.
Congregational Church, Jno Irwin, pastor.
DOHERTY, J H, saloon.
Jones, A, clothing.
Likins, F L, drugs & jewelry.
Lucette, R J, hardware.
MARKLEY & FAGAN, groceries.
McPhee, A, saloon.
Terrell, J K, clothing & shoes.
VINCENT, T W, lumber and coal.
Whitehead, J H, justice peace.
Wyland, Geo A, meat market.

1899[7]

BACHELOR (Teller, P. O.)

An important mining town in Mineral county, 2½ miles northwest of Creede. Population 200. Altitude 10,100.

Barnie & Co, A N, gen mdse.
Biles, J A, physician.
Congregational Church.
Jenkins, J W, saloon.

LIKENS, F L, drugs & jewelry.
Markley & Fagan, dry goods and groceries.
McDonald & McLeod, saloon.
O'Leary, Sim, saloon.
SLOAN, W C, lumber.
St. Cloud Hotel, Mrs P Daugherty, prop.
Terrell, J K, clothing.
Teter, J F, restaurant.
Vincent, T W, hardware.
Waters, C R, blacksmith.
Williams, W W, saloon.

1900[8]

BACHELOR (Teller, P. O.)

A mining town in Mineral county, 2½ miles northwest of Creede. Population 200. Altitude 10,100.

Barnie & Co. A N, gen mdse.
Biles, J A, physician.
Congregational Church.
HATCHER & WOOD, dry goods, hats and shoes.
Jenkins, J W, saloon.
Likens, F L, drugs & jewelry.
Markley & Fagan, dry goods and groceries.
O'Leary, Sim, saloon.
Sloan, W C, lumber.
St. Cloud Hotel, Mrs P Daugherty, prop.
Terrell, J K, clothing.
Teter, J F, restaurant.
Vincent, T W, hardware.
Waters, C R, blacksmith.
Williams, W W, saloon.

1901[9]

BACHELOR (Teller, P. O.)

A mining town in Mineral county, 2½ miles northwest of Creede. Population 200. Altitude 11,100.

Barnie & Co, A N, groceries.
Biles J A, physician.
Brotzman Sim, saloon.
Congregational Church.
Hatcher & Wood, dry goods.
Jenkins, J W, saloon.
LIKENS, F L, drugs and jewelry. hats and shoes.
Markley & Fagan, groceries.
Sloan, W C, lumber.

St. Cloud Hotel, Nellie Sullivan, prop.
Terrell, J K, clothing.
Teter, J F, restaurant.
Vincent, T W, hardware.
Waters, C R, blacksmith.

1902[10]

BACHELOR (Teller, P. O.)

A mining town in Mineral county, 2½ miles northwest of Creede. Population 200. Altitude 11,100.

Barnie & Co, A N, groceries.
Biles J A, physician.
Congregational Church.
Earle and Witt, saloon.
Fultz Ed, barber.
HATCHER & WOOD, dry goods.
Heckler C P, barber.
Jenkins J W, saloon.
LIKENS, F L, drugs & jewelry.
O'Loughlin J P, saloon.
Sloan, W C, lumber.
St. Cloud Hotel, Nellie Sullivan, prop.
Terrell, J K, clothing.
Teter, J F, groceries.
Vincent, T W, hardware.
Waters, C R, blacksmith.
Williams W W, saloon.
Witt Mrs J R, restaurant.
WOOD WILL J, postmaster.

1903[11]

BACHELOR (Teller, P. O.)

A mining town in Mineral county, 2½ miles northwest of Creede. Population 200. Altitude 11,100.

Congregational Church.
Earle F M, saloon.
Fultz Ed, barber.
Grace Ed, saloon.
Heckler, C P, barber.
Likens F L, drugs & jewelry.
Shellhamer & Soliner, saloon.
Sloan W C, lumber.
Smith J W, groceries.
Terrell J K, clothing.
Vincent T W, hardware.

Witt Mrs. J R, restaurant.
WOOD WILL J, postmaster, groceries.

1904[12]
BACHELOR (Teller, P. O.)

A mining town in Mineral county, 2½ miles northwest of Creede. Population 200. Altitude 11,100.

Connor Mrs Frank, restnt.
Eades A B, groceries.
Earle F M, saloon.
Fagan Mrs. restaurant.
Ferrell Nellie, restaurant.
Fultz Ed, barber.
Grace Ed, saloon.
Hatcher T W, dry goods.
Kaylor B B, drugs.
Shellhammer and Sheldon, saloon.
Vincent T W, hardware, fuel and feed.
WOOD WILL J, postmaster, groceries & meats.

1905[13]
BACHELOR (Teller, P. O.)

A mining town in Mineral county, 2½ miles northwest of Creede. Population 100. Altitude 11,100.

Connor Mrs Frank, restnt.
Eades A B, groceries.
Fagan Mrs, restaurant.
Fultz Ed, barber.
Kaylor B B, drugs.
Shellhammer Ed, saloon.
Vincent T W, hardware, fuel & feed.
WOOD WILL J, postmaster, groceries & meats, dry goods.
Woodmanse & Garnior, sln.

1906[14]
BACHELOR (Teller, P. O.)

A mining town in Mineral county, 2½ miles northwest of Creede. Population 200. Altitude 11,100.

Congregational Church.
Eades A B, groceries.

Fagan Mrs S E, restaurant.
Ferrell Mrs Dan, bdg house.
Fulst Edward, mayor, barbr.
Kaylor B B, drugs.
Lundy P M, city marshal, saloon.
McClintock G N, school tchr.
SHELHAMMER E J, prop Union saloon.
Vincent T W, hardware, coal and feed.
WOOD WILL J, postmaster, groceries and dry goods.

1907[15]
BACHELOR (Teller, P. O.)

A mining town in Mineral county, 2½ miles northwest of Creede. Population 100. Altitude 11,100.

Bell Frank, prin school.
Congregational Church.
Eades A B, groceries.
Fagan Mrs S E, restaurant.
Fulst Edward, mayor, barbr.
Kaylor B B, drugs.
LUNDY MRS EDITH, hardware and postmaster.
Lundy P M, genl mdse.
Shelhamer & Gardner, saln.

1908[16]
BACHELOR (Teller, P. O.)

A mining town in Mineral county, 2½ miles northwest of Creede. Population 100. Altitude 11,100.

Catholic Church, Father Good priest.
Congregational Church.
Eades A B, groceries.
Farrell Mrs Nellie, restrnt.
Fleck A V, saloon.
Fulst Edward, mayor, barbr.
Kaylor B B, drugs.
Lundy Mrs Annie, hardware.
LUNDY MRS EDITH, postmistress.
Lundy P M, genl mdse.
Moore & Gardner, saloon.
Sale L W, prin school.

1909[17]

BACHELOR (Teller, P. O.)

A mining town in Mineral county, 2½ miles northwest of Creede. Population 60. Altitude 11,100.

Catholic Church, Father Good priest.
Congregational Church.
Davis George, Saloon
Eades A B, groceries.
Fulst Edward, barber and mayor.
Lundy Miss Edith, postmistress.
Lundy P M, genl mdse. and hdw.
Moore & Boyson, saloon.

1910[18]

BACHELOR (Teller P. O.)

A mining town in Mineral county, 2½ miles northwest of Creede. Population 150. Altitude 10,100.

Davis & Co Chas H, saloon.
MOORE FRANK J, drugs and postmaster.

1915

There were no listings under Bachelor or Teller.

Endnotes

Chapter 1

1. The book was published by the Geological Society of America in 2000 as Special Paper 346. It is the official scientific report of geologic experiments conducted in the Creede area in the early 1990s and contains fourteen chapters written by various scientists who conducted the experiments. It was co-edited by Philip M. Bethke, a summer resident of Creede, since deceased. While the writing is technical, it is a great book for anyone interested in the history of Ancient Lake Creede.

2. The book was published in 2010. Chuck Harbert, the author, is a resident of Mineral County and a volunteer at the Creede Historical Society Museum. He is also a member of the museum committee. The appendix contains an easily readable discussion of the early geologic events, Ancient Lake Creede, the creation of mineral deposits, and the formation of the Upper Rio Grande Valley.

3. Peter W. Lipman, "Central San Juan Caldera Cluster: Regional Volcanic Framework," in *Ancient Lake Creede: Its Volcano-Tectonic Setting, History of Sedimentation, and Relation to Mineralization in the Creede Mining District,* Philip M. Bethke and Richard L. Hay, eds. Special Paper 346 (Boulder: Geological Society of America, 2000), 9.

4. Richard C. Huston, *A Silver Camp Called Creede: A Century of Mining,* Montrose, CO: Western Reflections, 2005), 7.

5. Lipman, "Central San Juan Caldera Cluster," 39.

6. Huston, *A Silver Camp Called Creede*, 8.

7. Paul B. Barton, Thomas A. Steven, and Daniel O. Hayba, "Hydrologic Budget of the Late Oligocene Lake Creede and the Evolution of the Upper Rio Grande Drainage System," in *Ancient Lake Creede: Its Volcano-Tectonic Setting, History of Sedimentation, and Relation to Mineralization in the Creede Mining District,* Philip M. Bethke and Richard L. Hay, eds., Special Paper 346 (Boulder: Geological Society of America, 2000), 108.

8. Lipman, "Central San Juan Caldera Cluster," 41. A color diagram can be found on page 62.

9. Barton, Steven, and Hayba, "Hydrologic Budget of the Late Oligocene Lake Creede," 105.

10. Huston, *A Silver Camp Called Creede*, 10.

11. The chapter by Barton, Steven, and Hayba has a diagram on page 109 showing the successive stages in the development of the Creede caldera.

12. Crater Lake in Oregon is an excellent example of a preserved caldera, lake, and resurgent dome.

13. Barton, Steven, and Hayba, "Hydrologic Budget of the Late Oligocene Lake Creede,"108.

14. Ibid.

15. Ibid.

16. Philip M. Bethke and Richard L. Hay, "Overview: Ancient Lake Creede," in Philip M. Bethke and Richard L. Hay, eds., *Ancient Lake Creede: Its Volcano-Tectonic Setting, History of Sedimentation, and Relation to Mineralization in the Creede Mining District.* Special Paper 346 (Boulder: Geological Society of America, 2000), 6–7; Paul B. Barton, Robert O. Rye, and Philip M. Bethke, "Evolution of the Creede Caldera and Its Relation to Mineralization in the Creede Mining District, Colorado," in Philip M. Bethke and Richard L. Hay, eds., Ancient Lake Creede: Its Volcano-Tectonic Setting, History of Sedimentation, and Relation to Mineralization in the Creede Mining District. Special Paper 346 (Boulder: Geological Society of America, 2000), 301–08.

17. Barton, Steven, and Hayba, "Hydrologic Budget of the Late Oligocene Lake Creede," 122–24.

18. Ibid., 108.

Chapter 2

1. Marilyn A. Martorano, Ted Hoefer III, Margaret A. Jodry, Vince Spero, and Melissa L. Taylor, *Colorado Prehistory: A Context for the Rio Grande Basin* (Denver: Colorado Council of Professional Archaeologists, 1999), 52.

2. Ibid., 49.

3. Ibid., 57.

4. O. Ned Eddins, *Prehistoric Migration of American Indians,* accessed August 15, 2016, www.thefurtrapper.com/prehistoricindians.hmtl.

5. Cathy E. Kindquist, *Stony Pass: The Tumbling and Impetuous Trail* (Silverton: San Juan County Book Company, 1987), 11.

6. Ibid., 11–14.

7. Ibid., 14.

8. Ibid., 16.

9. Nolie Mumey, *Creede: The History of a Colorado Silver Mining Town* (Denver: Artcraft: 1949), 5.

10. P. David Smith, *The Story of Lake City Colorado And Its Surrounding Areas* (Lake City, CO: Western Reflections, 2016), 28–29.

11. Ibid., 30–31.

12. Ibid., 33–34.

13. Kindquist, *Stony Pass,* 22.

14. Ibid., 23.

15. Ibid., 26.

16. Ibid., 2.

17. Ibid., 29.

18. Ibid., 43.

19. Ibid., 46.

20. Frances McCullough, "The Barlow and Sanderson Stage Line in the San Luis Valley," *San Luis Valley Historian,* 30, No. 3, 1998: 48–49.

21. Huston, *A Silver Camp Called Creede,* 19.

22. Janis Jacobs, *Ribs of Silver, Hearts of Gold,* vol. 2 (Creede: Creede Historical Society, 1994), 5.

23. Ibid., 6.

24. William H. Emmons and Esper S. Larsen, *Geology and Ore Deposits of the Creede District*, U.S. Geological Survey Bulletin 718 (Washington, DC: Government Printing Office, 1923), 3–4.

25. Charles W. Henderson, *Mining in Colorado*, U. S. Geological Survey Professional Paper 138, (Washington, D. C.: Government Printing Office), 1926, 56.

26. Huston, *A Silver Camp Called Creede*, 55.

27. Ibid., 36.

28. Ibid., Chapter 8 in Huston's *A Silver Camp Called Creede* has a detailed discussion of each of the mines on the Amethyst Vein.

Chapter 3

1. There are various spellings of MacKenzie in the literature, including MacKenzie, McKenzie, and Mackenzie. For example, the *Creede Candle* used McKenzie in its obituary and Mackenzie is used in *Mines and Mining Men of Colorado* (see bibliography). William H. Emmons and Esper S. Larsen in their U. S. Geological Survey Bulletin 718 (see bibliography) use MacKenzie. I have used the latter spelling throughout the book unless it is in a quote where a different spelling is used.

2. "Death of John McKenzie," *Creede (CO) Candle*, October 19, 1894, 1, col. 3.

3. *Teller Topics* (Bachelor, CO), July 22, 1892, 4, col. 1.

4. Richard S. Irwin, "Honor to Whom Honor Is Due," *Creede (CO) Candle*, January 13, 1893, 1, col. 3.

5. John G. Canfield, *Mines and Mining Men of Colorado: The Principal Producing Mines of Gold and Silver, The Bonanza Kings and Successful Prospectors, The Picturesque Camps and Thriving Cities* (Denver: John G. Canfield, 1893), 62.

6. "Death of John McKenzie," *Creede (CO) Candle*, October 19, 1894, 1, col. 3.

7. "Bachelor City: How and When It Was Started and by Whom Settled," *Colorado Sun* (Denver), September 11, 1892, 14.

8. Paul Gordon Bertelroot, "The History of Creede, a Mining Camp in Its Early Days" (Masters thesis, Denver University, 1953), 49.

9. Canfield, *Mines and Mining Men of Colorado*, 62.

10. "Around the Camp," *Creede Candle*, January 21, 1892, 4, col. 3.

11. Mumey, *Creede*, 157.

12. C. C. Davis, "Mines at Creede," *Herald Democrat* (Leadville, CO), March 2, 1892, 1, col. 1.

13. *Creede (CO) Candle*, April 8, 1892, 1, col. 3.

14. Ibid., "Notice," April 22, 1892, 2, col. 3.

15. Ibid., "Notice," June 24, 1892, 4, col. 3.

16. Ibid., "Around the Camp," April 22, 1892, 4, col. 2.

17. Duane Smith, *Henry M. Teller, Colorado's Grand Old Man* (Boulder: University Press of Colorado, 2002), 9–10.

18. "Around the Camp," *Creede (CO) Candle*, May 13, 1892, 4, col. 2.

19. *Teller Topics* (Bachelor, CO), July 22, 1892, 4, col. 1.

20. "Around the Camp," *Creede (CO) Candle*, June 24, 1892, 4, col. 3.

21. "City and County," *Lake City (CO) Times*, June 2, 1892, 4, col. 2.

Chapter 4

1. "Around the Camp," *Creede* (CO) *Candle*, April 22, 1892, 4, col. 2.
2. The author viewed and obtained copies of these at the Hinsdale County Museum in Lake City, Colorado, in September 2015. Nearly all of the documents were handwritten, and the plat was in its original blueprint format.
3. Ed O'Kelley is the person who murdered Bob Ford in Creede on June 8, 1892. Ford, in turn, was famous for shooting the outlaw Jesse James while the latter was standing on a chair fixing a picture. In the various incorporation documents, O'Kelley's last name is spelled O'Kelly, Kelley, and Kelly. The name O'Kelley is used because that is how he signed an April 13, 1892, document attesting to the population of Bachelor.
4. The signatures were in a handwriting style common in the late 1800s. The author did not enlist the aid of a handwriting expert. Rather, he did his best to translate the handwriting. As such, some misinterpretations are possible. The original copies are in the Hinsdale County (Colorado) Museum and Creede (Colorado) Historical Society Library to resolve any misinterpretation.
5. *Creede* (CO) *Candle*, June 3, 1892, 3, col. 4. The author could find no evidence that the election results were published in a Lake City newspaper.
6. Ibid., June 17, 1892, 4, col. 2.
7. *Lake City* (CO) *Times*, June 16, 1892, 4, col. 4.
8. *Creede* (CO) *Candle*, June 24, 1892, 1, col. 4.
9. Ibid., July 1, 1892, 1, col. 3.
10. Likens is spelled Likins in other documents.
11. "Council Proceedings," *Teller Topics* (Bachelor, CO), July 22, 1892, 1, col. 5.
12. Ibid., "Meeting of the Board of Trustees," July 30, 1892, 4, cols. 2–3.
13. Ibid., "Notice," July 30, 1892, 4, col. 3.

Chapter 5

1. Information from Bachelor City map (prepared by Field and McNutt and shown in fig. 24) in the Hinsdale County Museum, Lake City, CO.
2. The photograph was made from an original glass negative collection the author purchased in 2015.
3. This photograph was from the same collection described in note 2.
4. Susan Weston, *Family History of Harold Wheeler and Muriel LeZotte* (published by author, 1999), 16. The book is available at the Creede, CO, Historical Society Library.
5. Ibid., 18.
6. Ibid., 20–21.
7. *Teller Topics* (Bachelor, CO), July 22, 1892, 1, col. 4.
8. Mumey, Creede, 154; Huston, *A Silver Camp Called Creede*, 116. Both list the elevation as 10,500 feet above sea level.
9. *Creede* (CO) *Candle*, January 6, 1893, 2, col. 1.
10. The 334 total was a hand count by the author, taken from the U. S. Census information obtained at the Creede, CO, Historical Society Library.
11. Don LaFont, *Rugged Life in the Rockies*, 2nd ed., (Denver: Big Mountain, 1966), 65.
12. *Teller Topics* (Bachelor, CO), July 22, 1892, 4, col. 1.
13. Ibid., July 30, 1892, 4, col. 1.

14. Ibid., "Local Topics," October 1, 1892, 4.
15. Ibid., September 24, 1892, 4, col. 1.
16. "Around The Camp," *Creede* (CO) *Candle*, February 10, 1893, 4, col. 2.
17. Ibid., "Around The Camp," July 21, 1893, 4. Fouch and Foutch are used in the same article. It is spelled Foutch in the *1893 Colorado Business Directory*.
18. *Teller Topics* (Bachelor, CO), July 22, 1892, 4, col. 1.
19. Ibid., "Local Topics," 1, col. 4.
20. *American Newspaper Annual 1893–1894* (Philadelphia: N. W. Ayer and Sons), 64.
21. "Local Topics," *Teller Topics* (Bachelor, CO), July 30, 1892, 4, col. 1.
22. Ibid.
23. "Dies in a Chair," *Durango* (CO) *Democrat*, May 7, 1902, 1.
24. *Sacramento* (CA) *Daily Union*, June 5, 1894, 1, col. 3.
25. "Mines and Prospects," *Creede* (CO) *Candle*, January 13, 1893, 1, col. 3. The author was able to access only a four-month run of *Teller Topics* (July–October 1892) but he has been unable to locate copies of the *Sentinel* or *Tribune*, so it is unclear whether they were published weekly or monthly.
26. Accessed February 27, 2015, http//coloradowest.auraria.edu/newspapers-history/bachelor-teller.
27. "Mines and Prospects," *Creede* (CO) *Candle*, July 1, 1892, 1, cols. 2–3.
28. "The Spring Creek Road," *Teller Topics* (Bachelor, CO), September 10, 1892, 1, col. 3.
29. Ibid., September 17, 1892, 1, col. 3.
30. *Creede* (CO) *Candle*, September 16, 1892, 1, col. 2.
31. Ibid., "Mines and Prospects," September 23, 1892, 1, col. 2.
32. *Teller Topics* (Bachelor, CO), September 24, 1892, 1, col. 3.
33. Ibid., "Local Topics," October 1, 1892, 4, col. 2.
34. "Around the Camp," *Creede* (CO) *Candle*, January 21, 1892, 4, col. 3.
35. *Colorado Sun* (Denver), September 11, 1892, 14.
36. "Local Topics," *Teller Topics* (Bachelor, CO), July 22, 1892, 1, col. 4.
37. *Colorado Business Directory, 1893*, 138–141.
38. "Local Topics," *Teller Topics* (Bachelor, CO), July 30, 1892, 4, col. 2.
39. Ibid., August 6, 1892, 4, col. 3.
40. Ibid., "Local Topics," August 6, 1892, 4, col. 1.
41. Ibid., Sept. 10, 1892, 4, col. 1. Martindale's middle initial was "C.," not "O."
42. Ibid., July 22, 1892, 1, col. 5.
43. Ibid., 1, col. 3.
44. Information from Harold French Wheeler map and legend, figs. 5.6 and 5.7, respectively, and the 1904 Sanborn Fire Map of Bachelor.
45. Weston, *Family History*, 28.
46. "Local Topics," *Teller Topics* (Bachelor, CO), July 22, 1892, 1, col. 4.
47. Ibid., July 30, 1892, 4, col. 1.
48. Ibid., "Local Topics," August 13, 1892, 4, col. 2.
49. Ibid.
50. Information from Harold French Wheeler map and legend, figs. 5.6 and 5.7, respectively, and 1893 and 1904 Sanborn Fire Maps of Bachelor. The author has been unable to locate a photograph of the Catholic Church.
51. "Local Topics," *Teller Topics* (Bachelor, CO), July 22, 1892, 1, col. 5.
52. Ibid., "Local Topics," July 22, 1892, 1, col. 4.
53. Weston, *Family History*, 28.
54. "Local Topics," *Teller Topics* (Bachelor, CO), August 20, 1892, 4, col. 2.

55. Ibid., "Electric Lights," September 24, 1892, 4, col. 1.
56. "Electric Lights," *Creede* (CO) *Candle*, December 9, 1892, 4, col. 3.
57. Ibid., "Mines and Prospects," October 19, 1894, 1, col. 2.
58. Ibid., "To Haul Bachelor Ore," January 13, 1893, 4, col. 3.
59. Ibid., March 24, 1893, 4, col. 2.
60. "Want Better Roads," *Lake City* (CO) *Times*, September 8, 1892, 1, col. 4.
61. *Creede* (CO) *Candle*, "Around the Camp," December 16, 1892, 4, col. 2.
62. "A New County Wanted," *Lake City* (CO) *Times*, December 8, 1892, 1, col. 3.
63. *Creede* (CO) *Candle*, December 23, 1892, 4, col. 1.
64. "The City's Grist," *Lake City* (CO) *Times*, January 26, 1893, 4, col. 4.
65. "Passed the House," *Creede* (CO) *Candle*, March 24, 1893, 1, col. 2. Wason was named the county seat as a result of some political shenanigans by friends of Col. Martin Van Buren Wason.
66. Mumey, Creede, 92. The move of the county seat to Creede involved a bitter fight between Martin Van Buren Wason and the citizens of Creede. Details are given on pages 86–92.

Chapter 6

1. "Local Topics," *Teller Topics* (Bachelor, CO), July 22, 1892, 1, col. 4.
2. Ibid., "Local Topics," July 30, 1892, 4, col. 2.
3. Ibid., "In Memoriam," August 13, 1892, 4, col. 3.
4. Ibid., "Local Topics," August 13, 1892, 4, col. 2.
5. *Creede* (CO) *Candle*, April 8, 1892, 1, col. 3.
6. *Teller Topics* (Bachelor, CO), July 30, 1892, 4, col. 1.
7. Caroline Bancroft, *Unique Ghost Towns and Mountain Spots* (Boulder: Johnson, 1961), 75.
8. "Shooting at Bachelor," *Creede* (CO) *Candle*, June 24, 1892, 1, col. 4.
9. Ibid., July 1, 1892, 1, col. 3.
10. Ibid., "Around the Camp," July 29, 1892, 4, col. 2.
11. Ibid., "Around the Camp," June 17, 1892, 4, col. 3. McCoy had previously been arrested in June for an assault on Ed Sales.
12. *Teller Topics* (Bachelor, CO), July 30, 1892, 4, col. 1.
13. "A Killing at Bachelor," *Silver Standard* (Silver Plume, CO), April 1, 1893, 1, col. 2.
14. "Locked Out a Shift," *Creede* (CO) *Candle*, March 24, 1893, 1, col. 3.
15. Ibid., "Killing at Bachelor," March 24, 1893, 1, col. 4.
16. Ibid., "Around the Camp," March 3, 1893, 4, col. 3.
17. Ibid., "Killing at Bachelor," March 10, 1893, 1, col. 4. Mr. W. H. Woodruff's initials in this article were different from those in the first article (H. W.). They are typed as spelled in the respective articles. Which initials are correct is unknown.
18. *Teller Topics* (Bachelor, CO), September 17, 1892, 4, col. 1.
19. "Andy Wellington Killed," *Creede* (CO) *Candle*, July 22, 1905, 1, col. 3.
20. Ibid., "Around the Camp," June 17, 1892, 4, col. 2. A detailed account of the murder of Bob Ford can be found in Bob Ford by David Clark (see Bibliography). Clark was close to the correct name of the accomplice in Bob Ford: Jesse James' *Killer Shot Down in Creede* (see Bibliography). He states on page 26 that the *Creede Candle* identified the killer as "'Joe E_the_' (with the second and last letters of the last name unreadable)."

21. Ibid., May 5, 1893, 1, col. 3.
22. Edwin Lewis Bennett and Agnes Wright Spring, *Boom Town Boy: In Old Creede*, Colorado (Chicago: Sage Books 1966), 139.
23. Mumey, *Creede*, 159.
24. "Fires at Bachelor," *Creede* (CO) *Candle*, January 13, 1893, 4, col. 2.
25. Ibid., "Around the Camp," March 3, 1893, 4, col. 2.
26. Ibid., May 5, 1893, 1, col. 3.
27. Ibid., "Big Fire at Bachelor," January 26, 1894, 4, col. 2.
28. Ibid., "A Close Call in Bachelor," June 23, 1893, 1, col. 5.
29. Ibid., "Fire Started by A. Vance," February 17, 1893, 4, col. 3.
30. Ibid., "Around the Camp," March 24, 1893, 4, col. 2.
31. *Creede* (CO) *Chronicle*, May 13, 1893, 4, col. 1.
32. Ibid., May 13, 1893, 1, col. 1. I could find no further stories on whether Vance was recaptured and tried for attempted arson.
33. "Struck by Lightning," *Creede* (CO) *Candle,* June 15, 1912, 1, col. 4.
34. Weston, *Family History*, 16.
35. Ibid., 29.
36. It is easy for the author to identify with Mr. Wheeler's adventures and pranks because he grew up at the western edge of Pueblo, Colorado, with the same freedoms and opportunities for trouble as Mr. Wheeler. The author's experiences are told in *The Story of Charles*, a book he wrote for his grandchildren. His two grandsons in Creede are living similar experiences now.
37. Weston, *Family History*, 31.
38. Ibid., 31–32.
39. Bennett and Spring, *Boom Town Boy*, 137.
40. Ibid., 137–138.
41. Ibid., 142.
42. Weston, *Family History*, 34.
43. Ibid., 35.
44. Ibid.
45. Jack Foster, "Jeep Diary: A Great Town in Its Day," *Rocky Mountain News* (Denver), October 4, 1952, 25.
46. Weston, *Family History*, 35.
47. Dave Southworth, *Colorado Mining Camps*, (Jacksonville, FL: Wild Horse, 1997), 169.
48. "Ladies Aid Society Festival," *Teller Topics* (Bachelor, CO), August 20, 1892, 1, col. 3.
49. Ibid., September 3, 1892, 4, col. 1.
50. Ibid., "With Music and Dancing: Opening of the New Boarding House of the Last Chance a Social and Artistic Success," September 3, 1892, 4, col. 2.
51. "Entertainment at Bachelor," *Creede* (CO) *Candle*, December 16, 1892, 4 col. 3.
52. "Preserve the Shadow Ere the Substance Fades," *Teller Topics* (Bachelor, CO), August 20, 1892, 4, col. 2.
53. Ibid., "Local Topics," August 6, 1892, 4, col. 1.
54. Ibid., "Local Topics," August 13, 1892, 4, col. 2.
55. Ibid., 4, col. 1.
56. Ibid., 4, col. 2.
57. "One Good Turn Deserves Another," *Creede* (CO) *Candle*, June 29, 1894, 1, col. 3.
58. Ibid., "Around the Camp," June 29, 1894, 1, col. 2.
59. Ibid., 1, col. 3.

60. Ibid., "The Program, Bachelor's Grand Blowout on the Great and Glorious Fourth," June 29, 1894, 1, col. 3.
61. Ibid., June 22, 1894, 1, col. 4.
62. Ibid., "A Large Block of Granite," June 29, 1894, 1, col. 4.
63. Ibid., "The Big Day: Bachelor People Attended Well to Its Celebration in Mineral County," July 6, 1894, 1, col. 2.
64. Ibid., "The Chronicler's Notes," July 6, 1894, 1, col. 2.
65. Ibid., "The Horse Races," July 6, 1894, 1, cols. 1–2. James Workman owned Workman's Ranch, located twenty miles up the Rio Grande from Creede. It is now Freemon Ranch.

Chapter 7

1. Huston, *A Silver Camp Called Creede*, 45. Chapter 3 in this book discusses the issue of silver politics in detail.
2. Accessed August 31, 2016, "Bland-Allison Act," http://www.1878.u-s-history.com/pages/h718.hmtl.
3. Accessed August 31, 2016, "Sherman Silver Purchase Act", http://www.u-s-history.com/pages/h762.hmtl.
4. Huston, *A Silver Camp Called Creede*, 46.
5. Accessed August 31, 2016, "Sherman Silver Purchase Act", http://www.u-s-history.com/pages/h762.hmtl.
6. Accessed August 31, 2016, http://www.historycentral.com/Industrialage/Silver Act.hmtl.
7. Huston, *A Silver Camp Called Creede*, 47.
8. "Creede Mines Closing," *Denver Republican*, June 30, 1893, 2, col. 2.
9. "Mines and Prospects," Creede (CO) Candle, July 7, 1893, 1, col. 2.
10. While the Commodore Tunnel was also under construction at the same time, it is not discussed here because its impact on Bachelor was minimal compared to the Nelson-Wooster-Humphreys Tunnel because it serviced only one mine.
11. Ibid., 207.
12. Ibid., 218–219.
13. Built by Albert E. Humphreys, construction on the mill began in 1901 and was completed early in 1902.
14. Bennett and Spring, *Boom Town Boy*, 138.
15. *Colorado State Business Directories* from 1893-1910 and 1915 are reproduced in their entirety in Appendix 2.
16. Ibid., "School Program at Bachelor," March 19, 1910, 1, col. 1.
17. Ibid., "Local Siftings," May 27, 1911, 4, col. 2.
18. Ibid., "Hello! Hello," January 13, 1906, 1, col. 3.
19. Ibid., "Rays of Light," February 10, 1906, 8, col. 5.
20. Ibid., "Rays of Light," June 30, 1906, 8, col. 5.
21. Ibid., August 18, 1906, 1, col. 4.
22. Ibid., "Rays of Light," March 25, 1905, 8, col. 5.
23. Ibid., "Fire at Bachelor," December 2, 1911, 1, col. 1.
24. Ibid., "Fire Scare," June 25, 1910, 1, col. 5.
25. Ibid., "Mountain Climbing Autos," August 12, 1911, 3, col. 4.
26. Ibid., August 19, 1911, 4, col. 3.

27. Ibid., "Motors to Bachelor," June 24, 1916, 1, col. 1.
28. Ibid., "Notice to Voters," October 17, 1914, 4, col. 3.
29. Ibid., "Local Siftings," March 1, 1913, 4, col. 1.
30. Ibid., "Landmark Removed," March 8, 1913, 1, col. 2.
31. Muriel Sibell Wolle, *Stampede to Timberline: The Ghost Towns and Mining Camps of Colorado*, 2nd ed. (Chicago: Sage 1974), 331.
32. "Rays of Light," *Creede* (CO) *Candle*, September 15, 1906, 8, col. 1.
33. Ibid.
34. Ibid., "Rays of Light," August 11, 1906, 8, col. 5.
35. Ibid., November 2, 1907, 2, col. 3.
36. Ibid., "Rays of Light," June 23, 1906, 8, col. 1. His last name was spelled Shelhamer in a companion article.
37. Ibid., "Collections," August 29, 1908, 4, col. 3.
38. Ibid., "Additional Locals," May 2, 1908, 2, col. 2.
39. Ibid., January 28, 1911, 3, col. 6.
40. Ibid., "Local Siftings," December 18, 1909, 6, col. 2.
41. Ibid., 6, col. 1.
42. Ibid., "Local Siftings," March 6, 1909, 4, col. 1.
43. Ibid., "Local Siftings," October 5, 1912, 4, col. 1.
44. Ibid., "Local Siftings," September 13, 1913, 4, col. 2.
45. Ibid., "Bachelor: A Farewell Party," January 22, 1910, 4, col. 3.
46. Ibid., "Proceedings of Board of County Commissioners," October 25, 1913, 2, col. 1.
47. Ibid., "Delinquent Tax List for 1919," October 16, 1920, 4, col. 2.
48. Ibid., "Local Siftings," August 30, 1913, 4, col. 2.
49. Ibid., "Local Siftings," October 30, 1915, 4, col. 3.
50. Weston, *Family History*, 23–24.
51. James A. Walls, Letter to Creede Historical Society, September 30, 1994.
52. Interview with Mary Johnson, October 1, 2012. The story was confirmed by former Mineral County Sheriff Phil Leggitt.
53. Taken from informational poster at Creede, Colorado post office, summer 2015. There is reason to question some of the facts on the poster. For example, A. B. Gades is almost certainly Asa B. Eades (see Chapter 9). Will Wood was probably postmaster before October 1903 because he is listed as postmaster in the *1902 Colorado Business Directory*. Similarly, the spelling "Mary Cassedy" may be incorrect because she is listed in the *1897 Colorado Business Directory* as Mary J. Casey, postmistress (see Appendix 2).
54. "Local Siftings," *Creede* (CO) *Candle*, January 24, 1912, 4, col. 2.
55. The closure date is from William H. Bauer, James L. Ozment, and John H. Willard, *Colorado Postal History: The Post Offices* (Crete NE: J-B Publishing, 1971), 127.

Chapter 8

1. "A Magic City," *Saturday Evening Post*, December 23, 1927, 108–9.
2. Information from phone conversation with Norah Dooley Korn, October 27, 2015.
3. Delma Dooley brought the key to the Creede Historical Society Museum on September 26, 2015. She related much the same story about finding the key but asked me to call Norah Dooley Korn for specifics.

4. The author purchased the token from George Carpenter, a former Creede resident. His son purchased the token in Del Norte from an antique dealer. Information about the saloon was taken from a photocopy of the Bachelor license registration purchased with the token.
5. Jack Foster, "A Great Town in Its Day," *Rocky Mountain News* (Denver), October 4, 1952, 25. The shooting deaths of Andy Wellington and John Erskine are described in Chapter 6.
6. Wolle, *Stampede to Timberline*, 328. It is likely that Wolle took the road east to the Commodore Mine rather than continuing up the hill to Bachelor.
7. Ibid., 329.
8. Caroline Bancroft, *Unique Ghost Towns and Mountain Spots* (Boulder: Johnson, 1961), 76. The author is aware of the story of the three people buried in the same grave but he was unable to confirm it through primary references and, therefore, did not include it in the book.
9. Sheila Goodman, "What Became of Bachelor City," *Mineral County Miner* (Creede, CO), June 15, 2006, 1, 8A.
10. Information from several telephone conversations with George Carpenter in April and May 2016.
11. The Creede Museum was established in the 1940s. When the Creede Historical Society was formed in 1984, it took over management of the museum.

Chapter 9

1. Weston, *Family History*, 27.
2. "Rays of Light," *Creede (CO) Candle*, October 21, 1905, 8, col. 1.
3. Weston, *Family History*, 25.
4. Canfield, *Mines and Mining Men of Colorado*, 62–63.
5. Weston, *Family History*, 26–27.
6. Huston, *A Silver Camp Called Creede*, 121.
7. Information provided by Clyde Dooley, interview, June 2015.
8. Weston, *Family History*, 25.
9. Weston, *Family History*, 24–25.
10. Huston, *A Silver Camp Called Creede*, v.
11. Larry Gardanier, email, July 17, 2013.
12. Bennett and Spring, *Boom Town Boy*, 138.
13. Information provided in emails from Lewis (Bud) J. Wood on January 23, 2016, January 25, 2016 and May 2, 2016.

Appendix 1

1. Mr. Likens was not listed in the June 24, 1892, article in the *Creede Candle* announcing the newly elected trustees. He was apparently elected between June 24 and the July 19, 1892 meeting of the trustees.
2. "Council Proceedings," *Teller Topics* (Bachelor, CO), July 22, 1892, 1, col. 5.
3. Ibid., Series of 1892, no. 1, 4, cols. 2–3.
4. Ibid., Series of 1892, no. 2, 4, cols. 3–4.
5. Ibid., Series of 1892, no. 3, 5, cols. 1–2.
6. Ibid., Series of 1892, no. 4, 5, cols. 2–3.
7. Ibid., Series of 1892, no. 5, 5, col. 4.
8. Ibid., Series of 1892, no. 6, 5, col. 4.

9. Ibid., Series of 1892, no. 7, 5, col. 4.
10. Ibid., Series of 1892, no. 8, 5, cols. 4–5.
11. Ibid., Series of 1892, no. 9, 5, col. 5.
12. Ibid., Series of 1892, no. 10, 5, cols. 5–6.
13. Ibid., Series of 1892, no. 11, 5, col. 6.
14. Ibid., Series of 1892, no. 12, 5, col. 6.
15. Ibid., Series of 1892, no. 13, 5, col. 6.
16. Ibid., Series of 1892, no. 14, July 30, 1892, 4, col. 4. The original version of the ordinance was published in the July 22, 1892, issue of *Teller Topics*, but the word Creede was at the end of section 1. Creede was changed to Bachelor in the version re-published on July 30.
17. Ibid., Series of 1892, no. 15, July 22, 1892, 8, col. 6.
18. *Teller Topics (Bachelor, CO) July 30, 1892,* no. 15, August 6, 1892, 4, col. 4. This was the second no. 15 published in *Teller Topics*. The first, concerning accounts, was published on July 22, 1892 (see note 17). The second, concerning dogs, was published on August 6, 1892, as indicated. Whether this was intended to be ordinance 16 is not known.

Appendix 2

1. *Colorado Business Directory*, (Denver: Ives Publishing, 1893), 139–41.
2. *Colorado Business Directory*, (Denver: Colorado Directory Publishing, 1894), 149–50.
3. Ibid., 1895, 539–40.
4. *Colorado Business Directory*, (Denver: Gazzetter Publishing, 1896), 131–32.
5. Ibid., 1897, 136–37.
6. Ibid., 1898, 141. Earlier directories referred to Bachelor as the "center" of the Creede Camp. This issue lists Bachelor as "an important mining town" in Mineral County. It was changed to "a mining camp" in the 1900 directory.
7. Ibid., 1899, 139.
8. Ibid., 1900, 138.
9. Ibid., 1901, 140. This issue and subsequent issues list the elevation as 11,100 feet above sea level. Previous editions list the elevation as 10,100 feet above sea level.
10. Ibid., 1902, 143.
11. Ibid., 1903, 143–44.
12. Ibid., 1904, 144.
13. Ibid., 1905, 145–46.
14. Ibid., 1906, 142.
15. Ibid., 1907, 150.
16. Ibid., 1908, 153.
17. Ibid., 1909, 152.
18. Ibid., 1910, 151.

Bibliography
Books, Articles, and Theses

Bancroft, Caroline, *Unique Ghost Towns and Mountain Spots.* Boulder: Johnson, 1961.

Barton, Paul B., Robert O. Rye, and Philip M. Bethke. "Evolution of the Creede Caldera and Its Relation to Mineralization in the Creede Mining District, Colorado." In *Ancient Lake Creede: Its Volcano-Tectonic Setting, History of Sedimentation, and Relation to Mineralization in the Creede Mining District,* ed. Philip M. Bethke and Richard L. Hay. Special Paper 346. Boulder: Geological Society of America, 2000, 301–26.

Barton, Paul B., Thomas A. Steven, and Daniel O. Hayba. "Hydrologic Budget of the Late Oligocene Lake Creede and the Evolution of the Upper Rio Grande Drainage System." In *Ancient Lake Creede: Its Volcano-Tectonic Setting, History of Sedimentation, and Relation to Mineralization in the Creede Mining District.* ed. Philip M. Bethke and Richard L. Hay. Special Paper 346. Boulder: Geological Society of America, 2000, 105–26.

Bauer, William H., James L. Ozment, and John H. Willard. *Colorado Postal History: The Post Offices.* Crete NE: J-B Publishing, 1971.

Bennett, Edwin Lewis, and Agnes Wright Spring. *Boom Town Boy: In Old Creede, Colorado.* Chicago: Sage Books, 1966.

Bertelroot, Paul Gordon. "The History of Creede, a Mining Camp in Its Early Days," Masters thesis, Denver University, 1953.

Bethke, Philip M., and Richard L. Hay, eds. *Ancient Lake Creede: Its Volcano-Tectonic Setting, History of Sedimentation, and Relation to Mineralization in the Creede Mining District.* Special Paper 346. Boulder: Geological Society of America, 2000.

Bcthke, Philip M., and Richard L. Hay, "Overview: *Ancient Lake Creede.*" In Ancient Lake Creede: Its Volcano-Tectonic Setting, History of Sedimentation, and Relation to Mineralization in the Creede Mining District. ed. Philip M. Bethke and Richard L. Hay. Special Paper 346. Boulder: Geological Society of America, 2000, 1–8.

Brown, Robert L. *An Empire of Silver: An Illustrated History.* Denver: Sundance Publications, 1984.

Colorado Business Directory, Denver: Ives Publishing, 1893.

Colorado Business Directory, Denver: Colorado Directory Publishing, 1894–1895.

Colorado Business Directory, Denver: Gazzetter Publishing, 1896–1911, 1915. Gazzetter was spelled Gazzateer in the 1896 directory.

Canfield, John G. *Mines and Mining Men of Colorado: The Principal Producing Mines of Gold and Silver, the Bonanza Kings and Successful Prospectors, the Picturesque Camps and Thriving Cities.* Denver: John G. Canfield, 1893.

Eddins, O. Ned. *Prehistoric Migration of American Indians.* Accessed January 9, 2009. www.thefurtrapper.com/prehistoric_indians.html.

Emmons, William H., and Esper S. Larsen. *Geology and Ore Deposits of the Creede District, Colorado*. U. S. Geological Survey Bulletin 718. Washington, DC: Government Printing Office, 1923.

Harbert, Charles A. *Colorado History: Insights and Views through Postcards*. Wellington, CO: Vestige Press, 2006.

Harbert, Charles A. *Creede, Colorado History: Insights and Views through Postcards and Photographs*. Wellington, CO: Vestige Press, 2010.

Henderson, Charles W. *Mining in Colorado: A History of Discovery, Development and Production*. U. S. Geological Survey Professional Paper 138. Washington, DC: Government Printing Office, 1926.

Huston, Richard C. *A Silver Camp Called Creede: A Century of Mining*. Montrose, CO: Western Reflections, 2005.

Jacobs, Janis. *Ribs of Silver, Hearts of Gold*. Vol. 2. Creede, CO: Creede Historical Society, 1994.

LaFont, Don. *Rugged Life in the Rockies*, 2nd ed. Denver: Big Mountain, 1966.

Larsen, Esper S. *Recent Mining Developments in the Creede District, Colorado*. U.S. Geological Survey Bulletin 811-B. Washington, DC: Government Printing Office, 1929.

Kindquist, Cathy E. *Stony Pass: The Tumbling and Impetuous Trail*. Silverton, CO: San Juan County Book Company, 1987.

Lipman, Peter W. "Central San Juan Caldera Cluster: Regional Volcanic Framework." In *Ancient Lake Creede: Its Volcano-Tectonic Setting, History of Sedimentation, and Relation to Mineralization in the Creede Mining District*. ed. Philip M. Bethke and Richard L. Hay. Special Paper 346. Boulder: Geological Society of America, 2000, 6–69.

Mangan, Terry Wm. *Colorado on Glass: Colorado's First Half Century as Seen by the Camera*. Denver: Sundance, 1975.

Martorano, Marilyn A., Ted Hoefer III, Margaret A. Jodry, Vince Spero, and Melissa L. Taylor. *Colorado Prehistory: A Context for the Rio Grande Basin*. Denver: Colorado Council of Professional Archaeologists, 1999.

McCullough, Frances. "The Barlow and Sanderson Stage Line in the San Luis Valley." *San Luis Valley Historian* 30, no. 3. Alamosa, CO: San Luis Valley Historical Society, 1998, 4–66.

Mumey, Nolie. *Creede: The History of a Colorado Silver Mining Town*. Denver: Artcraft, 1949.

Pelton, A. R. *San Luis Valley Illustrated*. Reprint of 1891 ed. Alamosa, CO: Ye Olde Print Shop, 2005.

Smith, Duane. *Henry M. Teller, Colorado's Grand Old Man*. Boulder: University Press of Colorado, 2002.

Smith, P. David. *The Story of Lake City Colorado And Its Surrounding Areas*. Lake City, CO: Western Reflections, 2016.

Southworth, Dave. *Colorado Mining Camps*, Jacksonville, FL: Wild Horse, 1997.

Steven, Thomas A., and Peter W. Lipman. *Calderas of the San Juan Volcanic Field, Southwestern Colorado*. Geological Survey Professional Paper 958. Washington, DC: Government Printing Office, 1976.

Weston, Susan. *Family History of Harold Wheeler and Muriel LeZotte*, Published by author, 1999.

Wolle, Muriel Sibell. *Stampede to Timberline: The Ghost Towns and Mining Camps of Colorado*, 2nd ed. Chicago: Swallow, 1974.

Newspapers

American Newspaper Annual 1893–1894 (Philadelphia)
Colorado Sun (Denver)
Creede (CO) Candle
Creede (CO) Chronicle
Cripple Creek (CO) Herald
Denver Republican
Durango (CO) Democrat
Durango (CO) Herald
Herald Democrat (Leadville, CO)
Lake City Times (Lake City, CO)
Mineral County Miner (Creede, CO)
Monte Vista (CO) Graphic
Rocky Mountain News (Denver)
Sacramento (CA) Daily Union
Silver Standard (Silver Plume, CO)
Teller Topics (Bachelor, CO)

Magazines

Saturday Evening Post

Interviews

George Carpenter, various conversations, Spring 2016
Clyde Dooley, June 2016
Delma Dooley and Bill Dooley, September 26, 2015
Norah Dooley Korn, October 27, 2015
Mary Johnson, October 1, 2012

Museums/Libraries

Adams State College Library, Alamosa, CO
Creede Historical Society Museum
Denver Public Library, Western History Collection
Stephen H. Hart Library and Research Center, Denver
Hinsdale County Museum, Lake City, CO

Other

Wood, Mae, xiii, 134
Woodruff, W. H., 78–79
Wooster Tunnel, 98
Workman, James, 94

Index

Adams, Mr. and Mrs. William, 118–19
Alden, George, 12
Alden, Gustav, 12
Allen, Amanda, 134
Allen, Archie, 135
Allen, Arthur R., 81–82, 120, 134, 135
Allen, Mabel, 120, 134
Allen, Olive, 120, 134
Amethyst claim/mine, 14, 20, 105
Amethyst Vein, 13–16, 20
Ancestral Continental Divide, 7
Ancient Lake Creede, 1
Anderson, Chas. A., 33
Antelope Springs, 12
Apex Hotel/House, 61–62, 86
Arnold, Sam, 126
Atkinson, Henry, 101
Bachelor caldera, 2, 6
Bachelor Sentinel, 53
Bachelor City/town
    bank, 50–51
    Business Directory listings, 165–77
    business section, 44–47
    businesses, 55-63
    cemetery, 67
    Catholic Church, 64–65, 92–93
    Congregational Church, 65–66, 115, 142
    culture, 92–97.
    decline in people and businesses, 107-109
    distance from Creede, 44, 112
    electric lights, 67–68
    elevation, 48–49
    fires, 84–87, 111–12
    first child born in, 74
    founding, 18–24
    gambling in, 123
    incorporation, 29–37
    jail, 47, 64, 124
    July 4, 1894, celebration, 97–101
    map, 26, 43, 47
    Opera House, 95
    people leaving from, 115-20
    plat/survey, 23–24, 42–44
    population, 50, 108
    post office, 27–28, 121–22
    railroad to, 68–70
    relationship with Hinsdale County, 69–71
    remains, 127–33
    saloon (Brotzman & Long) token, 125
    school, 65–66, 108–09
    telephone service, 109–10
    town ordinances, 37–40, 146–64
    town hall, 64
    town seal, 38–39, 157-58
    violence in, 75–83
    voting precinct discontinued, 114–15
    water pump, 45–46
    wooden sidewalks, 46, 110, 128–29
Bachelor City Dramatic Club, 92, 95–96
Bachelor Mine, 20, 105
Bachelor Tribune, 53
Bailas, W. H., 33
Baker, Charles, 9
Bancroft, Carolyn, 75, 129
Barnard, Geo. S. 33
Barry and Zumiebel Orchestra, 97
B. B. Simmons mining claim, 25
Bennett, Edwin Lewis, vii, 89–91, 106, 138
Bennett, H. M., 14
Berry, "Scotty," 111, 115, 118, 134
Bethke, Philip, viii, 5
Big Ingun mining claim, 25
Biles, Dr. John A., 134
bimetallism, 102
Bland-Allison Act of 1878, 103
Boggs, Tom, 9
Bresser, Rena, vii
Brotzman, C. E., 124
Brown, John V. 33
Bruns, Miss Mabel, 118
Buddenbock, Eric, 14
Calvin/Colvin, C. L., 22, 33
Campbell, L. E., 14, 98
Cantwell, E. R., 89
Carnahan, Cheryl, vii
Carpenter, George, vii, 131
Carrell, W. H., 33
Carson, Kit, 9
Cassedy, Mary J., 121
Caywood, Miss Willie, 66
Central San Juan Caldera cluster, 1–2

Coffin, Sam, ii, 14, 22–23, 33, 134
Coinage Act of 1873, 102
Collins, James H., 99
Collins, Pat, 85
Colorado Business Directory, 165–77
Conway, Harry, 83
Cook, Geo, H., 33
Cooley, A. F., 114
Coyne, Thomas, 75–76
Crawford, A. S., 24, 33, 35, 37, 135
Creede caldera, 1–4, 7
Creede and Gunnison Shortline Railroad, 68–70
Creede Formation, 4, 5
Creede Historical Society Archives/ Museum/Library, ii, v, vii, 117
Creede, Nicholas/N.C., 14, 18
Davis, C. C., 24
David, Oscar, 109
Davis, George, 111
Davlin, Doug, vii, 133
Davlin, Toni, vii, 133
Dempsey, Jack, 135-137
Dillon, Richard, 33
Doering, Steve, 91
Donnelly, Mike, 76
Dooley, Alma Lucille Wintz, vii, 124
Dooley, Andy, vii, 136-137
Dooley, Bill, vii, 124
Dooley, Clyde, vii
Dooley, Delma, vii, 124
Downey, Father (catholic priest), 64
Driver, James, 99
Duckworth, Liz Morton, viii
Duncan, Charles, town marshal, 78
Duncan, W. C., 115
Eades, Asa B., 111, 137, 186
Either, Joseph, 31–33, 35, 50, 83, 137
Ellsberry and Foutch Banking and Merchantile Co., 51, 108
Equitable mining claim, 25, 26
Erskine, John, 78–79, 127
Fleming, Mike, 116
Folsom culture, 8
Ford, Bob, i, 82
Foster, Jack, 91, 126
Foster, Mat., 113
Free Coinage Hotel, 61, 62–63, 85
Freemon Ranch, 11

Fuller, Mr. C. L. and Mrs. Amy, 74
Fulst, Edward ("the Barber"), 33, 137, 142
Gades, A. B., 121
Gardanier, Irene, vi
Gardanier, Larry, vi, 139
Gardanier, Sutter, 138
Gardenheir, William, 76–77
Gardner/Gardiner, Creede undersheriff, 78–79, 86–87
Garautto, Geo., W., 33
Gilliard, W. V., 14
Gilmor, Sarah, vi
Goodman, Charles, 48, 61, 96, 138
Goodman, Sheila, 130
Gould, John, 28, 30, 33–35, 37, 40, 56, 121, 138
Granger, Ralph, 14
Gray, Johanna, v, vi
Greenback mining claim, 22, 25
Haase, Julius, 14, 16
Hall's Restaurant, 60
Hamilton, Major E. M., 10
Hannifin, John, 33
Happy Thought Mine, 20, 88–89, 105
Harbert, Kay, viii
Hare, William J, 13
Hargraves, Orrin (Junior), 91
Hawkins, P. M., 33
Hill, Miss Verna, 118
Hilton, E. F., 85, 96
Hinsdale County, 21, 29, 69, 71
Hinsdale County Museum, v, 30, 31, 45-46
Hoffman, Gustav, ii, 37, 39, 40, 59–60, 138
Hogue, Bill, 75–76
Holliday, Arthur B., 74
Houston, Grant, v, 39, 142
Hoyt, W. C., 33
Humphreys, Albert E., 105
Humphreys, Ira B., 113, 114
Humphreys Mill, 106
Humphreys Tunnel, 105
Huston, Richard (Dick) C., Dedication, ii, 16
Irwin, J. N. H., 14
Irwin, Richard S., 14, 18
Jackson, John R., 138
Jackson, William C., 138
Jackson, William T., Jr. 138
Jackson, William T. Sr., 138

Jacobs, Jan, v
Jenkins, J. W., 33, 35, 37, 40, 138
Johnson, Charles, 99
Johnson, Lute H., 36, 52
Johnson, Mary, vii
Jordan, Bill, 116
Kent, Professor A. R. 111
Kettle, Rev. J. B., 65
Korn, Norah Dooley, vii, 124–25
Ladies Aid Society, 92
LaFont, Don, vii, 50
La Garita caldera, 1,2
Lake Creede, 3, 4, 5
Lara, Frank H., 29, 33, 34, 35, 37
Last Chance claim/mine, i, 14, 16, 17, 94, 97, 104
Legal Tender mining claim, 25, 26
Leggitt, Phil, vii, 130
LeZotte, Muriel Belle, 140
Likens/Likens, F. L., 37, 40, 181
Little Penn mining claim, 25–26
Long, John, 30, 33–34, 37
Long, Lee, 124
Love, Ella, 61, 62
Love's News Depot, 61, 62
Lundy and Sherry saloon, 85
Lundy/Gardanier, Edith Eleanor Oberg, ii, vi, 121, 138–39
Lundy, Pat, 116, 117, 138
MacKenzie, John, ii, iii, 14, 16, 18–20, 22, 23, 139
MacMullen/McMullen, L., town marshal, 41, 77, 81
Maid of Erin mining claim, 25, 26
Marshall, Annie, 120
Marshall, Garrett (Gary), 120
Martindale/Martindale, George C., 23, 30, 33, 34, 139
Martorano, Marilyn, 8
McBride, David, 118
McCoy, "Kid," 76–77
McCullook, H., 33
McCullough, "Kid," 78–79
McCullough, Harry, 78
McGilvray, Colin, 99
McKune, Robert, 61, 62
McMullin, Harry, 99, 100
Miller, Cyrus, 112–113
Mineral County, creation 71–73
Moffat, David H., 14
Moore, Frank J., 121, 139
Moran, P. T., 33
Morton, Jane, iii, vii, 143–45
Mulhall, M. J., 33
Naylor, E. R. 14
Nelson, Charles F., 14, 105
Nelson Tunnel, 105–106
Nelson-Wooster-Humphreys Tunnel, 106
Newland, I. W., 33, 35, 37, 139
New York-Chance Mine, 17
O'Kelley, Ed, i, ii, 31–33, 82–83, 137, 140
Pacheco, Don Bernando de Miera y, 9
Palmer, General William Jackson, 13
Park Regent Mine, 15, 105
Parkinson, Bud., 87
Pennington, E. B., 33
Phipps, Genevieve, 113, 114
Phipps, Senator Lawrence, 114
Pierce, Carol, v
Pierson, C. H., 22, 37, 40, 55, 140
Pollack, Mrs. Laura, 66
Poker Alice, viii, 123–24
Quaking Asp mining claim, 22, 25, 26
Raney/Roney, F. B., 33
Regan, Jack, 99
Renniger, Theodore, 14, 16
Reynolds, Albert E., 77–78
Rice, E. S., 26
Ryden, Fred, 91, 126–28
Sail, Mr. 91
St. James Hotel, 61, 62
San Juan City, 11
San Juan Ranch, 11
Schneider, William, vii, 11
Seago, Bob, iv, v, vi
Shelheimer, Ed, 116
Sherman Silver Purchase Act of 1890, 103
Sherry, Michael, 80
Shorten, John, 51–53, 55, 57
Smith, George K., 14
Smith, G. L., 14
Smith, Jan, viii
Smith, P. David, viii
Snowshoe Mountain, 3, 4, 5
Soufa/Souva, F. L. sheriff, 35, 75–76
Soward, Jackson, 12
Spencer and Watson, attorneys, 29, 30
Spero, Vincent, 8

Spring Creek Pass Road, 53–54
Steel, Geo. W., 33
Stone, Miss Agnes, 96
Stony Pass, 10–11
Teller, Henry M., U. S. Senator, 27–28
Teller Topics, 51–53, 55–57
Thatcher, J., 85
Thatcher, Newt J., 23
Tubbs, Alice (see Poker Alice)
Uran, Andy, 118
Ute Indians, 8
Vance, A. A., 33, 86–87
Van Noorden, S. E., 121
Vincent, T. W., 24, 30, 33, 34, 35, 39, 56, 85,
    115, 140
Wason Ranch, 12
Wellington, Andy, 81–82, 127, 134
Weston, Susan, vii, 88
Wheeler, Harold French, vii, 45, 46, 47, 64,
    65, 66, 87–89, 140
Whelan, P. H., 33
Whitehead, A. H., 37, 40, 56, 60, 140
White mining claim, 25, 26
Wilson, Del, 89, 99
Wise, Mrs. J. L., 120
Winthrow, J. W., 96
Wolle, Muriel Sibell, 128–29
Wood, Lewis J., vii. 140–41
Wood, Mae, vii, 140
Wood, Will, vii, 116, 121, 140–41
Woodruff, W. H or H. W., 79–80
Wooster Tunnel, 105–106
Workman, James, 101

www.ingramcontent.com/pod-product-compliance
Lightning Source LLC
Chambersburg PA
CBHW070349090426
42733CB00009B/1349